Advanced Synchrotron Radiation Techniques for Nanostructured Materials

Advanced Synchrotron Radiation Techniques for Nanostructured Materials

Special Issue Editor

Chiara Battocchio

MDPI • Basel • Beijing • Wuhan • Barcelona • Belgrade

MDPI

Special Issue Editor
Chiara Battocchio
Roma Tre University
Italy

Editorial Office
MDPI
St. Alban-Anlage 66
4052 Basel, Switzerland

This is a reprint of articles from the Special Issue published online in the open access journal *Nanomaterials* (ISSN 2079-4991) from 2018 to 2019 (available at: https://www.mdpi.com/journal/nanomaterials/special_issues/SR_nano)

For citation purposes, cite each article independently as indicated on the article page online and as indicated below:

LastName, A.A.; LastName, B.B.; LastName, C.C. Article Title. *Journal Name* **Year**, *Article Number*, Page Range.

ISBN 978-3-03921-680-2 (Pbk)
ISBN 978-3-03921-681-9 (PDF)

Cover image courtesy of Christoph Giesen and Michael Heuken.

Contents

About the Special Issue Editor vii

Chiara Battocchio
Advanced Synchrotron Radiation Techniques for Nanostructured Materials
Reprinted from: *Nanomaterials* **2019**, *9*, 1279, doi:10.3390/nano9091279 . . . . **1**

Valeria Secchi, Stefano Franchi, Marta Santi, Alina Vladescu, Mariana Braic, Tomáš Skála,
Jaroslava Nováková, Monica Dettin, Annj Zamuner, Giovanna Iucci and Chiara Battocchio
Biocompatible Materials Based on Self-Assembling Peptides on Ti25Nb10Zr Alloy: Molecular
Structure and Organization Investigated by Synchrotron Radiation Induced Techniques
Reprinted from: *Nanomaterials* **2018**, *8*, 148, doi:10.3390/nano8030148 . . . . **4**

Yi-Ting Cheng, Hsien-Wen Wan, Chiu-Ping Cheng, Jueinai Kwo, Minghwei Hong and
Tun-Wen Pi
Microscopic Views of Atomic and Molecular Oxygen Bonding with *epi* Ge(001)-2 × 1 Studied
by High-Resolution Synchrotron Radiation Photoemission
Reprinted from: *Nanomaterials* **2019**, *9*, 554, doi:10.3390/nano9040554 . . . . **23**

Jaanus Kruusma, Arvo Tõnisoo, Rainer Pärna, Ergo Nõmmiste and Enn Lust
In Situ X-ray Photoelectron Spectroscopic and Electrochemical Studies of the Bromide Anions
Dissolved in 1-Ethyl-3-Methyl Imidazolium Tetrafluoroborate
Reprinted from: *Nanomaterials* **2019**, *9*, 304, doi:10.3390/nano9020304 . . . . **37**

Anton Plech, Bärbel Krause, Tilo Baumbach, Margarita Zakharova, Soizic Eon,
Caroline Girmen, Gernot Buth and Hartmut Bracht
Structural and Thermal Characterisation of Nanofilms by Time-Resolved X-ray Scattering
Reprinted from: *Nanomaterials* **2019**, *9*, 501, doi:10.3390/nano9040501 . . . . **64**

David Smrčka, Vít Procházka, Vlastimil Vrba and Marcel B. Miglierini
On the Formation of Nanocrystalline Grains in Metallic Glasses by Means of In-Situ Nuclear
Forward Scattering of Synchrotron Radiation
Reprinted from: *Nanomaterials* **2019**, *9*, 544, doi:10.3390/nano9040544 . . . . **81**

Xinghui Long, Pengfei Yu, Nian Zhang, Chun Li, Xuefei Feng, Guoxi Ren, Shun Zheng,
Jiamin Fu, Fangyi Cheng and Xiaosong Liu
Direct Spectroscopy for Probing the Critical Role of Partial Covalency in Oxygen Reduction
Reaction for Cobalt-Manganese Spinel Oxides
Reprinted from: *Nanomaterials* **2019**, *9*, 577, doi:10.3390/nano9040577 . . . . **92**

Eleonora Secco, Heruy Taddese Mengistu, Jaime Segura-Ruíz, Gema Martínez-Criado,
Alberto García-Cristóbal, Andrés Cantarero, Bartosz Foltynski, Hannes Behmenburg,
Christoph Giesen, Michael Heuken and Núria Garro
Elemental Distribution and Structural Characterization of GaN/InGaN Core-Shell Single
Nanowires by Hard X-ray Synchrotron Nanoprobes
Reprinted from: *Nanomaterials* **2019**, *9*, 691, doi:10.3390/nano9050691 . . . . **104**

Taddäus Schaffers, Thomas Feggeler, Santa Pile, Ralf Meckenstock, Martin Buchner,
Detlef Spoddig, Verena Ney, Michael Farle, Heiko Wende, Sebastian Wintz,
Markus Weigand, Hendrik Ohldag, Katharina Ollefs and Andreas Ney
Extracting the Dynamic Magnetic Contrast in Time-Resolved X-ray Transmission Microscopy
Reprinted from: *Nanomaterials* **2019**, *9*, 940, doi:10.3390/nano9070940 . . . . **115**

About the Special Issue Editor

Chiara Battocchio, Prof., is Associate Professor in General Chemistry at the Department of Sciences of the Roma Tre University, where she carries out research in the field of materials science. Her research activity is mainly focused on the study of the molecular and electronic structure of nanostructured materials as carried out by state-of-the-art techniques such as synchrotron radiation X-ray photoelectron spectroscopy (SR-XPS), X-ray absorption spectroscopy (XAS), and near-edge X-ray absorption fine structure spectroscopy (NEXAFS), as well as conventional XPS, UV—visible absorption and emission, and FTIR spectroscopy. As for the investigated materials, her attention is mainly devoted to organic molecules and biomolecules, anchored as self-assembling monolayers on macroscopic "flat" surfaces and/or used as capping agents for the stabilization of noble metal nanoparticles. When the macroscopic support is a biocompatible material such as, for example, titanium or titanium alloys, promising biocompatible materials are obtained for applications in the field of tissue engineering. Spectroscopic techniques such as SR-induced X-ray photoemission and absorption spectroscopies allow for probing the bio(organic)molecule/inorganic material interface, obtaining unique information about local structure as well as the molecular and electronic properties of the hybrid system.

nanomaterials

MDPI

Editorial

Advanced Synchrotron Radiation Techniques for Nanostructured Materials

Chiara Battocchio

Department of Science, Roma Tre University, 00146 Rome, Italy; chiara.battocchio@uniroma3.it;
Tel.: +39-065773-3400

Received: 3 September 2019; Accepted: 6 September 2019; Published: 7 September 2019

Nanostructured materials exploit physical phenomena and mechanisms that cannot be derived by simply scaling down the associated bulk structures and behaviors; furthermore, new quantum effects come into play in nanosystems. The exploitation of these emerging nanoscale interactions prompts the innovative design of nanomaterials.

Understanding the behavior of materials on all length scales, from the nanostructure up to the macroscopic response, is a critical challenge for materials science. Modern analytical technologies based on the synchrotron radiation (SR) allow the non-destructive investigation of the chemical, electronic, and magnetic structure of materials in any environment. The SR facilities have developed revolutionary new ideas and experimental set-ups to characterize nanomaterials, involving spectroscopy, diffraction, scatterings, microscopy, tomography, and all kinds of highly sophisticated combinations of such investigation techniques.

This Special Issue seeks to cover all aspects of synchrotron radiation applied to the investigation of chemical, electronic, and magnetic structure of nanostructured materials. It is composed by eight research articles, that together provide not only an interesting and multi-disciplinary overview on the chemico-physical investigation of nanostructured materials carried out by state-of-the-art synchrotron radiation induced techniques, but also an exciting glance on the future perspectives of nanomaterials characterization methods. The published papers focus on the chemical, structural, and morphological characterization of nanostructured materials of different nature carried out by a wide selection of SR-induced techniques such as the X-ray photoelectron spectroscopy (XPS) [1,2] and near edge X-ray absorption fine structure (NEXAFS) [1], in situ XPS and electrochemistry [3], time-resolved X-ray scattering [4], nuclear forward scattering (NFS) of synchrotron radiation [5], soft X-ray absorption spectroscopy (sXAS) [6], X-ray fluorescence [7], X-ray diffraction [7], X-ray absorption [7] and time-resolved X-ray transmission microscopy [8].

In the following, a brief overview of the individual articles published in this Special Issue will be provided, with the aim to elicit the interest of potential readers.

In the first paper of the Special Issue [1] V. Secchi et al. exploited an SR-induced XPS and angular dependent NEXAFS to investigate the chemisorption of a self-assembling peptide (EAbuK16, i.e., H-Abu-Glu-Abu-Glu-Abu-Lys-Abu-Lys-Abu-Glu-Abu-Glu-Abu-Lys-Abu-Lys-NH$_2$) onto annealed Ti25Nb10Zr alloy surfaces; the data acquired on Ti25Nb10Zr discs after incubation with self-assembling peptide solution at five different pH values allowed the investigation of the best conditions for peptide immobilization, by comparing the quality (coverage, molecular order) of the obtained nanostructured films. The X-ray photoemission spectroscopy at a high resolution, allowed by the use of SR as an X-ray source, was also exploited in the work from Y.-T. Cheng et al. [2] to investigate the embryonic stage of oxidation of an epi Ge(001)-2 × 1 by atomic oxygen and molecular O$_2$. The authors observed that the topmost buckled surface with the up- and down-dimer atoms, and the first subsurface layer behaved distinctly from the bulk by exhibiting surface core-level shifts in the Ge 3d core-level spectrum, and that O$_2$ molecules underwent dissociation upon reaching the epi Ge(001)-2 × 1 surface. The SR-induced XPS allowed the authors to observe that the down-dimer Ge atoms and the back-bonded subsurface atoms remained inert towards atomic O and molecular O_2, a behavior which might account for the low reliability in the Ge-related metal-oxide-semiconductor (MOS) devices. Still using XPS, but in situ experimental conditions in the field of electrochemistry, J. Kruusma et al. [3] investigated the influence of electrode potential on the electrochemical behavior of a 1-ethyl-3-methylimidazolium tetrafluoroborate (EMImBF$_4$) solution containing 5 wt% 1-ethyl-3-methylimidazolium bromide (EMImBr). The authors followed the evolution of the Br $3d_{5/2}$ XPS signal, collected in a 5 wt% EMImBr solution at an EMImBF$_4$–vacuum interface, and were able to detect the start of the electrooxidation process of the Br^- anion to Br^{3-} anion and thereafter to the Br_2 at the micro-mesoporous carbon electrode, polarized continuously at the high fixed positive potentials. Moreover, the B1s spectral region was monitored allowing to evidence B–O bond formation at E ≤ −1.17 V, parallel to the start of the electroreduction of the residual water at the micro-mesoporous carbon electrode. This study demonstrates the excellent potentiality of in situ SR-induced spectroscopies in investigating the details of electrochemical processes in operando. In situ observations of chemico-physical phenomena are nowadays allowed by several techniques; in this context, D. Smrčka et al. [5] report about the application of nuclear forward scattering (NFS) of synchrotron radiation to the in situ study of crystallization of metallic glasses. By performing in situ temperature experiments in the presence and absence of an applied magnetic field, they are able to carry on the investigation not only from the structural point of view, evidencing the formation of nanocrystalline grains, but also to observe the evolution of the corresponding hyperfine interactions, and the differences in the NFS spectrum evolution induced by the experimental conditions (i.e., temperature and magnetic field). Temperature-related effects on the nanostructured materials behavior are also studied by A. Plech et al. [4], which propose an innovative method to investigate the thermal conductivity of thin films exploiting the high time-resolution in scattering analysis. In their manuscript, the authors apply transient pump-probe detection of dissipation of laser-induced heating (TDXTS) to investigate two extreme examples of phononic barriers, isotopically modulated silicon multilayers with very small acoustic impedance mismatch and silicon-molybdenum multilayers, which show a high resistivity; the reliable results obtained and described allow the validation of the proposed method. In [6], X. Long et al. focus their attention on the structural investigation of nanocrystalline multivalent metal spinels carried out by sXAS; the object of this research is to identify the active sites in cubic and tetragonal $Co_xMn_{3-x}O_4$ (x = 1, 1.5, 2) spinel oxides (a family of highly active catalysts for the oxygen reduction reaction—ORR) and to understand their reaction mechanisms, since these aspects are essential to explore novel transition metal oxides catalysts and further promote their catalytic efficiency. The authors demonstrate that the ORR activity for oxide catalysts primarily correlates to the partial covalency between the O 2p orbital with Mn^{4+} 3d t_{2g}-down/e_g-up, Mn^{3+} 3d e_g-up and Co^{3+} 3d e_g-up orbitals in octahedron, which can be directly revealed by the O K-edge sXAS. The findings reported in this publication highlight the importance of electronic structure in controlling the oxide catalytic activity. The X-ray absorption apectroscopy is also used by E. Secco et al. [7] to study individual nanowires (NWs) containing non-polar GaN/InGaN multi-quantum-wells, in a multi-technique investigation carried out by hard X-ray spectroscopies: X-ray fluorescence, X-ray diffraction, and X-ray absorption. Thanks to the improvements in the spatial resolution of synchrotron-based X-ray probes, that have now reached the nano-scale, the authors were able to probe the chemical composition of the nanowires as well as to describe the nanomaterial structure observing that while the GaN core and barriers appear fully relaxed, there is an induced strain in InGaN layers corresponding to a perfect lattice matching with the GaN core. Such strain, together with the observed inhomogeneous alloy distribution, affects the photoluminescence spectrum of non-polar InGaN quantum wells but still exhibits a reasonable 20% relative internal quantum efficiency. This study evidences how a multi-technique approach, allowed by the synchrotron radiation facilities, allows for a wide and accurate description of highly complicated materials, such as the ones proposed by the authors. The most recent publication in the Special Issue is the paper from T. Schaffers et al. [8] describing the application of a time-resolved detection scheme in the scanning transmission X-ray microscopy (STXM) to measure the element resolved ferromagnetic resonance (FMR) at microwave frequencies up to 10 GHz and a spatial resolution down to 20 nm at two different synchrotrons. The authors discuss different methods to separate the contribution of the background from the dynamic magnetic contrast based on the X-ray magnetic circular dichroism (XMCD) effect, and describe how the relative phase between the GHz microwave excitation and the X-ray pulses generated by the synchrotron, as well as the opening angle of the precession at FMR, can be quantified. In conclusion, the authors demonstrated how the dynamic magnetic contrast in time-resolved STXM has the potential to be a powerful tool to study the linear and nonlinear, magnetic excitations in magnetic micro- and nano-structures with unique spatial-temporal resolution in combination with element selectivity.

To conclude this overview on the papers published in the Special Issue "Advanced Synchrotron Radiation Techniques for Nanostructured Materials", I am confident that the readers will enjoy these contributions and may be able to find inspiration for their research within this Special Issue.

Funding: This research received no external funding.

Acknowledgments: The Grant of Excellence Departments, MIUR-Italy (ARTICOLO 1, COMMI 314–337 LEGGE 232/2016) is gratefully acknowledged. A special thank you to all the reviewers participating in the peer-review process of the submitted manuscripts and for enhancing their quality and impact; I am also grateful to Sandra Ma and the editorial assistants who made the entire Special Issue creation a smooth and efficient process.

Conflicts of Interest: The author declares no conflict of interest.

References

1. Secchi, V.; Franchi, S.; Santi, M.; Vladescu, A.; Braic, M.; Skála, T.; Nováková, J.; Dettin, M.; Zamuner, A.; Iucci, G.; et al. Biocompatible Materials Based on Self-Assembling Peptides on Ti25Nb10Zr Alloy: Molecular Structure and Organization Investigated by Synchrotron Radiation Induced Techniques. *Nanomaterials* **2018**, *83*, 148. [CrossRef] [PubMed]

2. Cheng, Y.-T.; Wan, H.-W.; Cheng, C.-P.; Kwo, J.; Hong, M.; Pi, T.-W. Microscopic Views of Atomic and Molecular Oxygen Bonding with epiGe(001)-2 × 1 Studied by High-Resolution Synchrotron Radiation Photoemission. *Nanomaterials* **2019**, *9*, 554. [CrossRef] [PubMed]

3. Kruusma, J.; Tõnisoo, A.; Pärna, R.; Nõmmiste, E.; Lust, E. In Situ X-ray Photoelectron Spectroscopic and Electrochemical Studies of the Bromide Anions Dissolved in 1-Ethyl-3-Methyl Imidazolium Tetrafluoroborate. *Nanomaterials* **2019**, *9*, 304. [CrossRef] [PubMed]

4. Plech, A.; Krause, B.; Baumbach, T.; Zacharova, M.; Eon, S.; Girmen, C.; Buth, G.; Bracht, H. Structural and Thermal Characterisation of Nanofilms by Time-Resolved X-ray Scattering. *Nanomaterials* **2019**, *9*, 501. [CrossRef] [PubMed]

5. Smrčka, D.; Procházka, V.; Vrba, V.; Miglierini, M.B. On the Formation of Nanocrystalline Grains in Metallic Glasses by Means of In-Situ Nuclear Forward Scattering of Synchrotron Radiation. *Nanomaterials* **2019**, *9*, 544. [CrossRef] [PubMed]

6. Long, X.; Yu, P.; Zhang, N.; Li, C.; Feng, X.; Ren, G.; Zheng, S.; Fu, J.; Cheng, F.; Liu, X. Direct Spectroscopy for Probing the Critical Role of Partial Covalency in Oxygen Reduction Reaction for Cobalt-Manganese Spinel Oxides. *Nanomaterials* **2019**, *9*, 577. [CrossRef] [PubMed]

7. Secco, E.; Mengistu, T.; Segura-Ruíz, J.; Martínez-Criado, G.; García-Cristóbal, A.; Cantarero, A.; Foltynski, B.; Behmenburg, H.; Giesen, C.; Heuken, M.; et al. Elemental Distribution and Structural Characterization of GaN/InGaN Core-Shell Single Nanowires by Hard X-ray Synchrotron Nanoprobes. *Nanomaterials* **2019**, *9*, 691. [CrossRef] [PubMed]

8. Schaffers, T.; Feggeler, T.; Pile, S.; Meckenstock, R.; Buchner, M.; Spodding, D.; Nev, V.; Farle, M.; Wende, H.; Wintz, S.; et al. Extracting the Dynamic Magnetic Contrast in Time-Resolved X-Ray Transmission Microscopy. *Nanomaterials* **2019**, *9*, 940. [CrossRef] [PubMed]

nanomaterials

MDPI

Article

Biocompatible Materials Based on Self-Assembling Peptides on Ti25Nb10Zr Alloy: Molecular Structure and Organization Investigated by Synchrotron Radiation Induced Techniques

Valeria Secchi [1], Stefano Franchi [1,*,†], Marta Santi [1], Alina Vladescu [2], Mariana Braic [2], Tomáš Skála [3], Jaroslava Nováková [3], Monica Dettin [4], Annj Zamuner [4], Giovanna Iucci [1] and Chiara Battocchio [1,*]

[1] Department of Science, Roma Tre University of Rome, Via della Vasca Navale 79, 00146 Rome, Italy; valeria.secchi@uniroma3.it (V.S.); santimarta3@gmail.com (M.S.); giovanna.iucci@uniroma3.it (G.I.)
[2] National Institute for Optoelectronics, 409 Atomistilor St., 077125 Magurele, Romania; alinava@inoe.ro (A.V.); mariana.braic@inoe.ro (M.B.)
[3] Department of Surface and Plasma Science, Faculty of Mathematics and Physics, Charles University, V Holešovičkách 2, 18000 Prague, Czech Republic; tomas.skala@elettra.eu (T.S.); jaroslava.lavkova@gmail.com (J.N.)
[4] Department of Industrial Engineering, University of Padua, Via Marzolo, 9, Padua 35131, Italy; monica.dettin@unipd.it (M.D.); annj.zamuner@studenti.unipd.it (A.Z.)
* Correspondence: stefano.franchi@elettra.eu (S.F.); chiara.battocchio@uniroma3.it (C.B.); Tel.: +39-040-3758059 (S.F.); +39-06-5733-3400 (C.B.)
† Present Address: Elettra-Sincrotrone Trieste S.C.p.A. di interesse nazionale, Strada Statale 14-km 163,5 in AREA Science Park, 34149 Basovizza, Trieste, Italy.

Received: 31 January 2018; Accepted: 5 March 2018; Published: 7 March 2018

Abstract: In this work, we applied advanced Synchrotron Radiation (SR) induced techniques to the study of the chemisorption of the Self Assembling Peptide EAbuK16, i.e., H-Abu-Glu-Abu-Glu-Abu-Lys-Abu-Lys-Abu-Glu-Abu-Glu-Abu-Lys-Abu-Lys-NH$_2$ that is able to spontaneously aggregate in anti-parallel β-sheet conformation, onto annealed Ti25Nb10Zr alloy surfaces. This synthetic amphiphilic oligopeptide is a good candidate to mimic extracellular matrix for bone prosthesis, since its β-sheets stack onto each other in a multilayer oriented nanostructure with internal pores of 5–200 nm size. To prepare the biomimetic material, Ti25Nb10Zr discs were treated with aqueous solutions of EAbuK16 at different pH values. Here we present the results achieved by performing SR-induced X-ray Photoelectron Spectroscopy (SR-XPS), angle-dependent Near Edge X-ray Absorption Fine Structure (NEXAFS) spectroscopy, FESEM and AFM imaging on Ti25Nb10Zr discs after incubation with self-assembling peptide solution at five different pH values, selected deliberately to investigate the best conditions for peptide immobilization.

Keywords: synchrotron radiation induced spectroscopies; XPS; NEXAFS; nanostructures; titanium alloy; self-assembling peptides; bioactive materials

---

1. Introduction

The increased interest in titanium (Ti) and its alloys for dental implants and prosthesis application derives from their exceptional mechanical properties, corrosion resistance and biocompatibility [1,2]. A gentle surgical technique combined with sufficient healing time has long been considered the key to osseo-integration, and excellent long-term clinical outcome for dental implant thus validates the results of pre-clinical experimental studies. Alloying improves the mechanical properties of titanium for use in high load-bearing applications, total hip, and total knee replacements. However, some concerns related to the toxicity of various alloying elements do exist [3]. In particular, Ti6Al4V alloy is commonly used in clinical practice as biocompatible material for prosthetics applications and dental implants [4–9]. In the last years, the in vitro and in vivo tests performed on Ti6Al4V alloy showed that this alloy has a toxic effect resulting from released V and Al and that its elastic modulus is very distant from the bone value [10–16], restricting its use in biomaterial applications. On this basis, a lot of experiments have been carried out to develop a novel Ti based alloy consisting only of biocompatible elements, which could replace the Ti6Al4V alloy in clinical practice [9,17–19]. For example, Ti6Al7Nb (ASTM F1295), Ti13Nb13Zr (ASTM F1713), and Ti12Mo6Zr (ASTM F1813) were proposed as candidates for manufacturing surgical implants. It is worth mentioning that there are a lot of other proposed alloys in the literature, such as Ti-Nb-Zr-Ta [20–24], Ti-Mo-Zr-Fe [25,26], Ti-Al-Zr [27], Ti-Al-Fe [18], Ti-Nb-Fe [28,29], Ti-Nb-Zr-Sn [30] and Ti-Nb [31] systems, but no standards have been published.

The here reported Ti-Zr-Nb system was selected for the following reasons: all of the constituent elements are considered to be highly biocompatible [32–34] and show high affinity to oxygen, leading to the formation of stable oxides which improve the corrosion resistance [35–39]; moreover, Zr is added in the alloy due to its capacity to stabilize the β phase. In fact, Abdel-Hady et al. showed that a Zr content ranging from 6 to 30 at % stabilized the β phase in alloys [40]. In the literature, there are few papers dealing with the effect of Zr content on the mechanical, tribological, and anticorrosive properties of Ti-Zr-Nb systems used for biomedical applications. In all published literature, the Zr content is up to 10 at % [41–46] or of about 30 at % [43]. Nb addition is also required because it maintains the β phase formed during the annealing. Furthermore a possible strategy to promote osseo-integration and enhance the biological acceptance of the implants is the biofunctionalization of the Ti25Nb10Zr surface with bioactive molecules that can be grafted on the surface in order to establish a molecular dialogue with host cells [47]. Among other bioactive molecules, self-assembling peptides (SAPs) are extremely promising candidates, since thanks to their on-purpose designed sequence they are able to self-assemble in a beta-sheet secondary structure [48,49]. They can then aggregate in the presence of saline creating hydrogels that can be used either as drug delivery vehicles, in the case of factors to release with a precise kinetic, or can be decorated with adhesive sequences or proteins, appropriately conjugated with a self-assembling sequence, allowing the functionalization of the scaffold with adhesive signals in a 3D structure by simple co-aggregation. The chemically stable SAP adhesion to the substrate is usually obtained by covalently and selectively functionalizing the alloy surface of the ion complementary peptide [50,51].

In this work, we present the characterization, carried out by synchrotron radiation-induced X-ray Photoemission Spectroscopy (SR-XPS), angle-dependent Near Edge X-rays Absorption Fine Structure (NEXAFS) spectroscopy, Field Emission Scanning Electron Microscopy (FE-SEM) and Atomic Force Microscopy (AFM) investigations of Ti25Nb10Zr alloy surfaces functionalized by the SAP EAbuK16 (Abu stands for α-aminobutyric acid), i.e., H-Abu-Glu-Abu-Glu-Abu-Lys-Abu-Lys-Abu-Glu-Abu-Glu-Abu-Lys-Abu-Lys-NH$_2$. The proposed SAP is able to self-assemble in aqueous solution in the presence of monovalent cations. To prepare the material, Ti25Nb10Zr discs were exposed to self-assembling peptide solutions at pH values ranging from 2 to 12, in order to understand the best conditions for peptide immobilization.

2. Materials and Methods

2.1. Samples Preparation

2.1.1. Ti25Nb10Zr Alloy Preparation and Preliminary Characterization

Ti25Nb10Zr (in wt %) was manufactured by Romanian Company (R&D Consulting and Services, Bucharest, Romania). Ti25Nb10Zr alloy was casted by a cold crucible levitation melting technique (CCLM), using a FIVES CELES—CELLES MP 25 furnace with nominal power 25 kW (Fives Celes, Lautenbach, France). The alloy was produced by mixing ultra-pure raw metals, subsequently annealed at 900 °C for 5 h in an oven (Caloris-CD 1121) (Caloris, Bucharest, Romania) and cooled in air. For this study, the alloy was cut as discs by a turning machine and mid-polished using 3 μm diamond emery paste.

The elemental composition and distribution of each constituent element of the alloy was checked by means of energy-dispersive X-ray spectroscopy (EDS), using the X-ray detector (EDS-Quantax70, Bruker, Billerica, MA, USA) attached to a scanning electron microscope (SEM, Hitachi TM3030PLUS) (Hitachi, Tokyo, Japan). The compositional analysis was performed automatically by the Quantax 70 microanalysis software (Bruker). The surface morphology of the Ti25Nb10Zr alloy substrates was investigated by atomic force microscopy (AFM) and scanning electron microscopy (SEM). AFM measurements were performed in tapping mode on 30×30 μm^2 area using an INNOVA microscope (Veeco, Plainview, NY, USA). The crystallographic structure was analyzed by X-ray diffraction (XRD) (Rigaku, Tokyo, Japan) using a diffractometer SmartLab Rigaku in the 2θ range 20–100°. The step Δ2θ was 0.02, and the minimum speed was 0.0002 deg/min. The CuKα radiation was used with a wavelength of λ = 1.5411 Å at 45 kV high voltage and 200 mA of the X-ray tube.

Since the alloy is prepared with the aim to be used as material for orthopaedic implants, special attention was devoted to the evaluation of the corrosion resistance in two solutions mimicking the physiological conditions: simulated body solution (SBF, composition: 8.035 g/L NaCl, 0.335 g/L NaHCO$_3$, 0.225 g/L KCl, 0.231 g/L K$_2$HPO$_4$·3H$_2$O, 0.311 g/L MgCl$_2$·6H$_2$O, 0.292 g/L CaCl$_2$, 0.072 g/L Na$_2$SO$_4$, 6.228 g/L Tris-(HOCH$_2$)$_3$CNH$_2$ [52]) and Hank solution (composition: 8 g/L NaCl, 0.4 g/L KCl, 0.1 g/L MgCl$_2$·6H$_2$O, 0.14 g/L CaCl$_2$, 1 g/L glucose, 0.35 g/L NaHCO$_3$, 0.06 g/L NaH$_2$PO$_4$·6H$_2$O, 0.06 g/L KH$_2$PO$_4$, 0.06 g/L MgSO$_4$ [53]). The corrosion resistance was evaluated by potentiodynamic polarization tests at 37 ± 0.4 °C using a VersaSTAT 3 Potentiostat/Galvanostat (Princeton Applied Research-AMETEK, Oak Ridge, TN, USA), following the steps:

Monitoring the open circuit potential (OCP) for 15 h after the immersion in electrolyte.

Plotting potentiodynamic curves −2 V vs. OCP to 2 V vs. SCE.

A conventional three-electrode cell was used, with a saturated calomel electrode (SCE) as reference, a platinum one as counter electrode, and the sample (1 cm^2) as working electrode. For the tests, the scanning rate was of 1 mV/s, value recommended also by ASTM G 59–97. During the tests, the solution was agitated by magnetic stirrer at 150 rpm for elimination of the gas bubbles formed during the test.

On the basis of potentiodynamic curves, both the corrosion potential ($E_{i=0}$) and corrosion current density (i_{corr}) were estimated. The polarization resistance (R_p) was calculated using the Stern–Geary Equation (1) [54]:

$$R_p = \frac{1}{2.3 \cdot i_{corr}} \cdot \frac{b_a \cdot b_c}{(b_a + b_c)} \tag{1}$$

where i_{corr} is corrosion current density, b_a is anodic slope and b_c is cathodic slope of the alloy.

The corrosion rates (CR) were calculated, on the basis of the values of the electrochemical parameters determined from the polarization curve, using Equation (2) according to the ISO G102-89 standard, reapproved in 1999:

$$CR = \frac{K \cdot i_{corr} \cdot EW}{\rho} \tag{2}$$

where K is a constant for units conversion, i_{corr}, the corrosion current density of the alloy (μA/cm^2 or A/cm^2), EW alloy equivalent weight (gram/equivalent), ρ alloy density (gram/cm^3).

2.1.2. Ti25Nb10Zr Alloy Surfaces Functionalization

SAP EAbuK16 was synthesized on solid phase as reported in [49].

Substrates were incubated for 18 h in aqueous solution at different pHs containing 1 mg/mL of EAbuK16. The SAP was dissolved in 10 mM NaCl (Carlo Erba, Cornaredo, Italy) aqueous solutions having different pH values: 0.1 mM HCl (J. T. Baker, Phillisburg, NJ, USA) (pH 4), 0.01 M HCl (pH 2), 0.1 mM NaOH (Carlo Erba) (pH 10), 0.01 M NaOH (pH 12). The pH 7 solution was prepared in two different ways: (a) buffered by Hank's solution (146.15 mg NaCl; 50 mg KCl (Carlo Erba); 287 mg Na_2HPO_4 (Carlo Erba); 50 mg KH_2PO_4 (Carlo Erba) in 250 mL distilled water); (b) 10 mM NaCl in distilled water.

The five solutions were then used to cover the alloy surfaces with a layer of SAP. More in detail, thin and thick self-assembling peptide layers namely monolayers, (MLs), and multilayers (MULs), were supported onto Ti25Nb10Zr surfaces as follows:

- MLs: Ti25Nb10Zr discs were sonicated in acetone for 5 min, dried, incubated in the peptide solution for 18 h, washed three times with NaCl 0.10 M at pH 7 and finally three times with distilled water. In these samples the set pH 7 solution was buffered with Hank's solution to mimic the extracellular physiological environment. Unfortunately, Hank's solution altered the ionic strength and interfered with peptide deposition. For this reason, the pH 7 sample was prepared again, avoiding the addition of sodium phosphate and other salts, except NaCl 10 mM, and maintaining the 10 mM NaCl washing treatment.

- MULs: Ti25Nb10Zr discs were sonicated in acetone for 5 min and dried. Peptide films were cast by covering the alloy surface with 2–3 drops of 1 mg/mL solutions of EAbuK16 oligopeptide prepared at different pHs, then dried in a low vacuum glass line.

2.2. Spectroscopic Techniques

2.2.1. X-ray Photoelectron Spectroscopy

High Resolution X-ray Photoelectron Spectroscopy (XPS) measurements on pristine alloy and multilayer samples were performed at the Materials Science Beamline (MSB) at the Elettra synchrotron radiation source (Trieste, Italy). MSB, placed at the left end of bending magnet 6.1, is equipped with a plane grating monochromator that provides light in the energy range of 21–1000 eV. The UHV endstation, with a base pressure of 1×10^{-8} Pa is equipped with a Specs Phoibos 150 hemispherical electron analyzer. Photoelectrons emitted by C1s, N1s, O1s, and Ti2p were detected at normal emission geometry using photon energy of 630 eV, estimated Energy Resolution = 0.6 eV. Binding energies were reported after correction for charging using the aliphatic C1s as a reference (B.E. 285.0 eV). Core-level spectra were fitted with a Shirley background and Gaussian peak functions.

Monolayer samples were investigated at the PM4-LowDosePES beamline at Helmholtz-Zentrum Berlin (BessyII Synchrotron Radiation facility), allowing for a lower flux on the sample, mandatory to avoid damaging the extremely thin layers of peptides. This soft X-ray bending magnet beamline is equipped with a Plane Grating Monochromator operating in collimated light (collimated PGM). It has two permanent end-stations, the reflectometer and the SURICAT (photoemission and X-ray absorption spectroscopy), which are in alternative operation. The LowDose PES end-station is equipped with an SES100 hemispherical analyzer [55]. Energy Resolution was estimated as 0.2 eV.

Conventional XPS studies were performed with an instrument designed by us, consisting of preparation and analysis chambers separated by a gate valve. The analysis chamber is equipped with a six-degree-of freedom manipulator and a 150-mm-mean radius hemispherical electron analyzer with a five-lens output system combined with a 16-channel detector. Ti2p, Nb3d, Zr3d, C1s, O1s, and N1s core level signals were recorded on the investigated samples unmonochromatized MgKα radiation; at least two specimens were analyzed for each sample type. Experimental spectra were analyzed by curve fitting using Gaussian curves as fitting functions; the analyzed spectra were energy referenced to the C1s signal of aliphatic–aromatic C–C carbons located at a binding energy B.E. = 285.0 eV [56,57]. Atomic ratios were calculated from peak areas using Scofield's cross as section sensitivity factors [58].

2.2.2. Near Edge X-ray Absorption Fine Structure Spectroscopy

Near Edge X-ray Absorption Fine Structure (NEXAFS) spectroscopy experiments were performed at the ELETTRA storage ring at the BEAR (bending magnet for emission absorption and reflectivity) beamline, installed at the left exit of the 8.1 bending magnet exit. The apparatus is based on a bending magnet as a source, a beamline optics delivering photons from 5 eV up to about 1600 eV with a selectable degree of ellipticity. The UHV end station has a movable hemispherical electron analyzer and a set of photodiodes to collect angle resolved photoemission spectra, optical reflectivity, and fluorescence yield, respectively. Moreover, it is equipped with ammeters in order to measure the total electron yield from the sample for NEXAFS measurements [59]. The carbon and nitrogen K-edge spectra were collected at normal (90°), magic (54.7°), and grazing (20°) incidence angles of the linearly polarized photon beam with respect to the sample surface. The photon energy and resolution (Energy Resolution: C K-edge 0.13 eV; N K-edge 0.2 eV) were calibrated and experimentally tested at the K absorption edges of Ar, N_2, and Ne. In addition, our carbon K-edge spectra were further calibrated using the resonance at 285.50 eV assigned to the C1s π^* ring transition. The spectra were then normalized subtracting a straight line that fits the part of the spectrum below the edge and assessing to 1 the value at 320.00 eV and 425.00 eV for carbon and nitrogen, respectively [48].

2.3. Microscopy Techniques

2.3.1. Field Emission Scanning Electron Microscopy

(FE-SEM) imaging studies on functionalized Ti25Nb10Zr surfaces were performed at the Charles University (Prague, Czech Republic) as preliminary investigation by means of a HITACHI S-4800 field emission scanning electron microscope operating at 30 keV electron beam energy.

2.3.2. Atomic Force Microscopy

AFM images were recorded on functionalized Ti25Nb10Zr surfaces using an INNOVA microscope (Veeco, Plainview, NY, USA) operating in tapping mode. Each image was acquired on 512 lines with 0.3 Hz on 30×30 μm^2 area. The SPMLab analysis software (Veeco) was used for data processing.

3. Results

3.1. Characterization of Pristine Ti25Nb10Zr

3.1.1. Assessment of the Elemental Composition and Homogeneity

The composition of the pristine alloy surface was probed by EDS analysis; in Table S1, presented in Supplementary material, the elemental composition determined in four different zones of the alloy surface is shown. It presents the following composition: Ti 68.8 wt %, Nb 21.9 wt %, and Zr 9.3 wt %. No significant differences were observed among the areas, indicating that each element is homogenously distributed on the whole surface. The small differences in EDS alloy composition compared to the value provided by the manufacturer are due to the uncertainty of the investigative techniques. The manufacturing company determined the chemical composition by spark emission spectroscopy. In addition, Figure S1 in the Supporting Information shows the EDS mapping images of the Ti25Nb10Zr alloy performed on the surface corresponding to zone 4 of Table S1. It is evident that each element is evenly distributed over the investigated surface, indicating that the constituent elements are homogenously mixed.

3.1.2. Crystallographic Structure

The crystallographic structure of the pristine Ti25Nb10Zr alloy was ascertained by means of XRD measurements. The XRD profile is shown in Figure 1. The reflection peaks from both α'' (orthorhombic) and β (disordered body-centered cubic) phases were detected. Phases identifications were performed by matching each peak with the JCPDS files No. 44-1284 (α'' phase) and 44-1288 (β phase)—Figure 1. Some planes of α'' and β phases were overlapped. The α'' phase is observed to be the predominant phase. Taking into account the plane of α'' phase located at 34.2°, 40.7°, and 52.2°, the grain sizes, calculated by the Scherrer formula, are about 18.7 nm, 15.9 nm, and 18.5 nm respectively, leading to an average of 17.7 nm. In the case of the β phase, taking into account the planes located at 55.7° and 95.4°, which are not overlapped on the α″ phase, the grain sizes were calculated to be 8.4 nm and 3.9 nm, with an average of 6.1 nm.

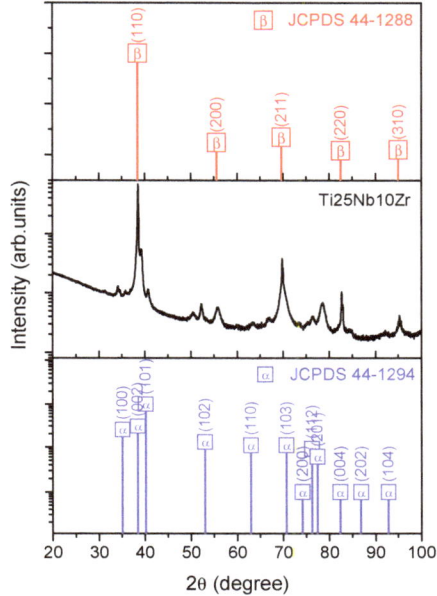

Figure 1. XRD diffraction pattern of Ti25Nb10Zr alloy and JCPDS files No. 44-1294 (α″ phase) and 44-1288 (β phase).

It is noteworthy that no diffraction peak of ω phase (hexagonal structure) was detected. This result is an advantageous one for the proposed alloy, because it is reported in the literature that ω phase causes embrittlement, leading to deterioration of mechanical properties [60–62].

3.1.3. Electrochemical Tests

Titanium has a good corrosion resistance in various corrosive environments such as seawater, organic chemicals, oxidizing or reducing acids over a wide range of concentrations and temperatures. This effect is due to oxide forming spontaneously and instantly when its surface is exposed to air and/or corrosive solutions [42]. Ishii et al. reported that the addition of a small amount (up to 3%) of alloying elements in the Ti matrix has a minor effect on the corrosion resistance of Ti in normally passive environments, while under active condition a small amount of alloying elements accelerate the corrosion process, leading to a significant deterioration of the alloy [42]. Thus, when we want to produce novel Ti-based alloys, the effect on its anticorrosive properties of the addition of a new alloying element in the basic matrix is the most important issue in determining whether the Ti passivity is lost and the surface has become fully active.

In the field of the design and production of novel materials for biomedical applications, the high corrosion resistance in physiological solutions is one the most important requirements, because it improves biocompatibility. Thus, for the present study, the Ti25Nb10Zr alloy was tested in terms of its corrosion resistance in two solutions: SBF and Hank at 37 ± 0.4 °C.

The evolution of the open circuit potential (E_{OCP}) during 15 h of immersion in SBF and Hank solutions at 37 °C is presented in Figure 2a. It is well known that a positive value of E_{OCP} indicates that the surface is coved with a protective oxide. In our study, one may see that the alloy immersed in

SBF exhibited a more negative value of E_{OCP} than that of alloy in Hank solution, but it was stabilized after 4 h of immersion. The evolution of E_{OCP} of Ti25Nb10Zr alloy immersed in Hank solution showed many fluctuations, indicating that the oxide is formed and destroyed due to the dissolution and re-passivation processes. This behavior is usually observed for metallic materials immersed in aggressive solutions, due to the presence of aggressive chloride or sulfate ions. Ti25Nb10Zr alloy tested in Hank solution showed some sudden falls of E_{OCP}, which can be attributed to the metastable pitting corrosion, apparently due to the pit anodic growth. After these falls, the E_{OCP} increased, indicating that the surface is re-passivated, by a cathodic oxygen-reduction reaction. This effect was also observed and reported in the literature by Isaacs et al. [63]. So, taking into account the E_{OCP} values, we can summarize that the Ti25Nb10Zr alloy is less affected by SBF attack.

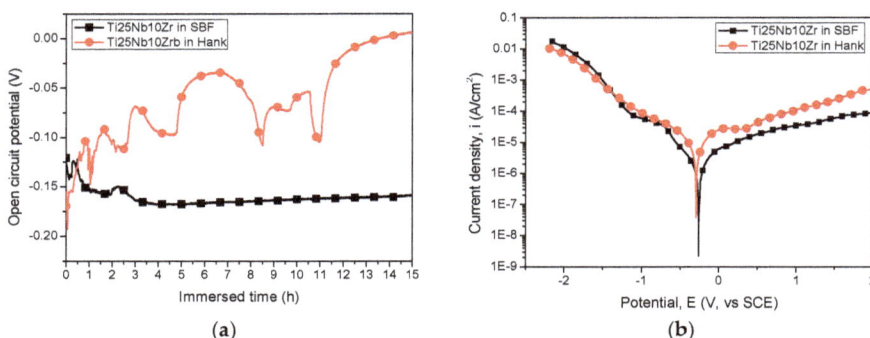

Figure 2. (**a**) Open circuit potential curves of Ti25Nb10Zr alloy in simulated body fluid (SBF) and Hank solutions (symbols are used only for identifying the curves in grey scale printing); (**b**) potentiodynamic polarization curves of Ti25Nb10Zr alloy in SBF and Hank solutions.

The potentiodynamic curves of the Ti25Nb10Zr alloy tested in SBF and Hank solutions at 37 °C are presented in Figure 2b. Based on these curves, the electrochemical parameters were determined, as summarized in Table 1. It is commonly accepted that a surface with a more electropositive corrosion potential, low corrosion current density, and high polarization resistance shows high corrosion behavior. Based on this statement, one may note that Ti25Nb10Zr alloy is more resistant to SBF corrosive attack. This result is also in good agreement with the corrosion rate results. A low corrosion rate was found for the alloy tested in SBF, indicating again that the alloy exhibited good resistance to SBF attack.

Table 1. The electrochemical parameters of Ti25Nb10Zr alloy after tests in simulated body fluid (SBF) and Hank solutions (E_{OCP}—open circuit potential; R_p—polarization resistance; E_{corr}—corrosion potential at $i = 0$; i_{corr}—corrosion current density; CR—corrosion rate). The E_{OCP} values were the last measured values in Figure 2a.

Electrolyte	E_{OC} (mV)	E_{corr} (mV)	i_{corr} (μA/cm^2)	R_p (Ω)	CR (mm/year)
SBF	−159	−253	1.15	45,011	0.011
Hank	6	−288	14.33	10,676	0.134

In the literature, it is reported that the corrosion rate is an inversely proportional relationship with pH, the corrosion rate decreasing when the pH increases [64]. In our study, the pH of solutions before corrosion tests was 7.4 for both solutions. After the corrosion tests, the pH values were 7.8 and 8.4 for SBF and Hank, respectively, indicating that the alloy is affected more or less by both solutions. In vivo, there is the possibility that the pH exceeds 7.8 in the vicinity of the implant, leading to an alkaline poisoning effect; all the physiological reactions will therefore be unbalanced and a more intense corrosion process will take place with a fast hydrogen evolution process [65]. On this basis, it can be expected that the alloy does not affect the tissue in the vicinity of implant in vivo because is more resistant in SBF. It is important to mention that the Ti25Nb10Zr alloy demonstrated good viability and proliferation after five days of culture with MG63 cells, having values approximately 10% higher than pure Ti [64].

Figure 3 shows SEM micrographs of the Ti25Nb10Zr alloy before and after corrosion tests at different magnifications. The corrosion tests showed a visible indication of deterioration of alloy in both solutions. No cracks, pores or macro-segregations can be observed on the alloy surface before corrosion (Figure 3a). Some defects can be observed on the alloy surface, probably generated during the polishing process. In both corrosive solutions, the Ti25Nb10Zr alloy suffers a localized corrosion; a relatively uniform attack over the exposed surface of the alloy was found. After corrosion, on the alloy surfaces, pits and cracks can be seen which are the main damage to the surface. Under the Hank testing solution, the alloy corroded more severely compared with alloy tested in SBF solution. Many pits are observed on the alloy tested in Hank solution compared with the SBF one. The localized corrosion increased the current density. According to SEM images, the surface tested in Hank solution is more affected, leading to an increase of current density. This finding is in good agreement with the above potentiodynamic curves. Some grey deposits are found on the alloy surfaces tested in both solutions, which can be related to corrosion products.

The SEM analysis is in good accordance with the open circuit potential measurements and polarization tests.

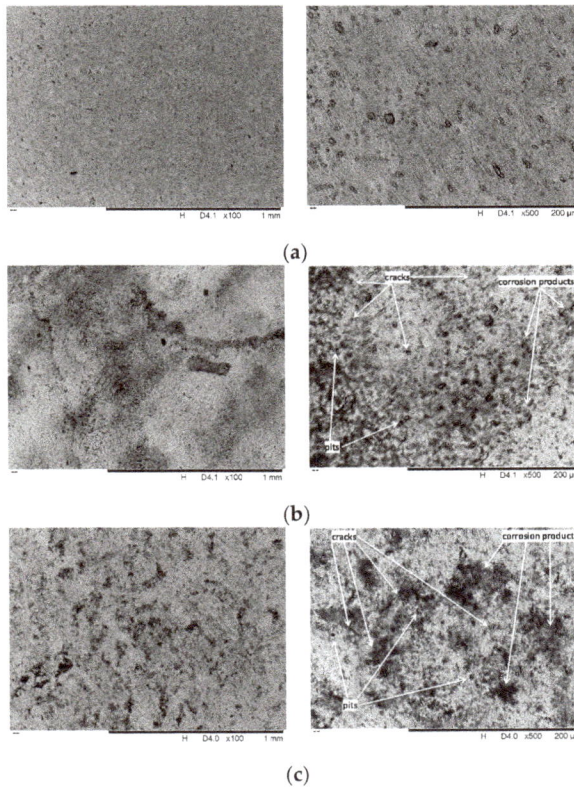

(a)

(b)

(c)

Figure 3. Scanning Electron Microscopy (SEM) images of the alloy surface before and after electrochemical tests. (**a**) Before corrosion; (**b**) after corrosion in SBF; (**c**) after corrosion in Hank.

3.1.4. Surface Composition

Pristine alloy discs were investigated by both Mg K-α source and SR-induced X-ray Photoelectron Spectroscopy (XPS) and with the aim to ascertain the chemical composition at the substrate surface; core level spectra of metal components, i.e., Ti, Nb, and Zr. Ti2p, Nb3d, and Zr3d signals were acquired, confirming the presence of the expected species TiO_2 (B.E. = 458.35 eV), Nb_2O_5 (B.E. = 206.98 eV) and ZrO_2 (B.E. = 182.10 eV) [56]. The observed atomic ratios Ti 75.4%, Zr 9.4%, Nb 15.2% do not exactly correspond to the expected stoichiometry due to the XPS sensitivity to the sample surface. In fact, since XPS is a surface-sensitive technique, the stoichiometry revealed refers to the outermost layers of the sample surface. Disparity with respect to the expected value could be due to phenomena of interdiffusion towards the outward layers.

Since pristine alloy XPS spectra were used as reference for the data analysis of the functionalized samples, O1s, C1s, and N1s core level signal, that are the most indicative for evaluating peptide grafting efficiency at different pH conditions, were also acquired. The O1s signal, reported in Figure 4a, shows a complex structure and, by following a peak-fitting procedure, at least three spectral components can be identified. The first peak at about 530 eV B.E. is attributed to the different metal oxides TiO_2, Nb_2O_5, and ZrO_2, which we were not able to discriminate, and therefore appear all grouped. At higher BE values a feature due to organic oxygen, arising by contaminating organic matter on the clean alloy surfaces (O–org, B.E. ~ 532 eV) can be singled out; a last component of lower intensity arising by physisorbed water is observed at about 533 eV B.E., as expected from literature for similar samples [65,66]. C1s spectrum is reported in Figure 4b. The curve fitting shows the presence of surface-contaminating carbons, leading to at least four contributions identified as aliphatic (C–C, B.E. = 285,0 eV), single-bonded to oxygen (C–O, B.E. = 286.23 eV), carbonyl-like (C=O, B.E. = 287.63 eV), and carboxyl-like (COOH, B.E. = 288.94 eV) carbons respectively. As expected, the N1s signal is not detected on the surface of "clean" samples.

On the other hand, a complete collection of B.E., Full Width Half Maximum (FWHM), Atomic Ratio values and Feature Assignments for the clean sample is reported in Table S2 in the Supporting Information.

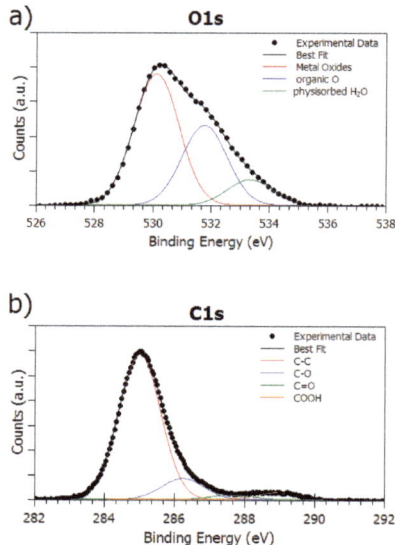

Figure 4. (a) XPS O1s and (b) XPS C1s spectra of the pristine sample surface.

3.2. Characterization of Ti25Nb10Zr Surfaces Functionalized with EAbuK16

3.2.1. Analysis of SAP Multilayer (MUL) Adsorbed at Different pH Values

Samples prepared as MULs from EAbuK16 solutions at pH 2, 4, 7, 10, and 12 were analyzed with both Mg K-α source and SR-induced radiation with the aim to probe molecular stability and anchoring efficiency at different pH values. As shown in Figure 5a for the sample prepared at pH 2 (taken as example), all samples show complex C1s signals (Figure 5a, top spectrum) which were analyzed by the curve-fitting procedure. The main component at lower B.E. values is attributed to aliphatic carbon atoms (C–C, B.E. = 285 eV) and was always used as calibration signal; the peak at about 286.3 eV BE is due to carbons bonded to nitrogens (as expected for aminoacids alpha-carbons, and lysine lateral groups) and/or oxygens; the third component is indicative of peptidic carbons N–C=O (amide-like, B.E. ~ 288.3 eV). The last feature, that can be observed as a small shoulder at higher BE values, is due to glutamic acid carboxyl groups (COOH, B.E. ~ 289 eV) [50,57]. C atoms due to contaminants are also observed in samples functionalized with peptides, giving rise to a signal that is superimposed to the peptide-one, but their amount is comparable in all investigated samples, since they were all treated by following the same procedure.

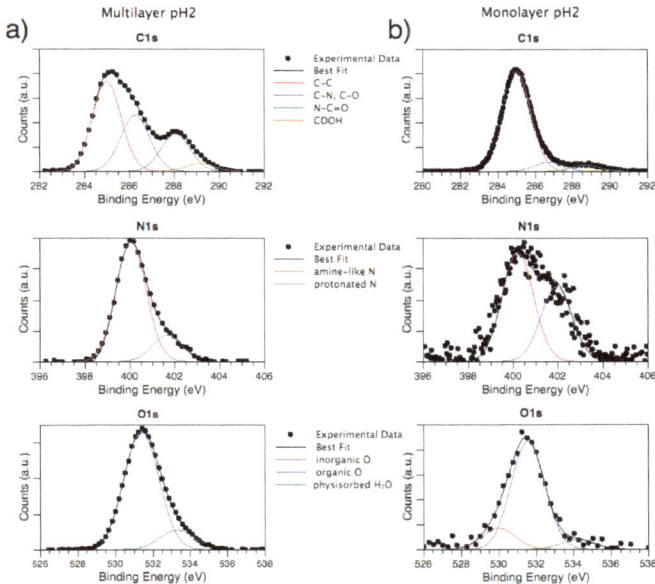

Figure 5. (a) X-ray Photoemission Spectroscopy (XPS) C1s, N1s, O1s MUL sample pH 2; (b) XPS C1s, N1s, O1s ML sample pH 2.

A complete collection of B.E. and FWHM values is reported in Tables S3 and S4 of Supporting Information. Quantitative estimations of different functional groups are reported in the following Table 2.

By exposing metal substrates to peptide solution, the signals related to nitrogen appear. More in detail, we can discriminate between protonated (B.E. ~ 402 eV) and non-protonated nitrogens (B.E. ~ 400 eV), as expected for peptides layers on metal surfaces, as shown in Figure 5a (spectrum in the middle) [50,57] O1s signals (Figure 5a, bottom) were also collected and compared with the ones observed for the clean substrate. Different from pristine alloy spectra, where all three oxygen species were detected, for MUL samples the signal related to metal oxides is absent, except in pH 7 and pH 12 samples. This indicates that in these two samples the peptide layer is thinner than in others, and since photoelectrons coming from metals are also detected, it is possible to estimate the peptide overlayer thickness by calculating the $Ti2p_{3/2}$ signal's intensity attenuation in samples with peptide, with respect to the signal intensity in a clean sample, according to equation 3.13 in [58].

Table 2. Atomic ratios of the species present on the multilayer (MUL) Self Assembling Peptide (SAP) surface. N^+: protonated nitrogen; Me–O_x: oxygen of metal oxides; O–org: organic oxygen.

MUL Sample	C–N/C–C Ratio	N–C=O/C–C Ratio	COOH/C–C Ratio	N_{tot}/C_{tot} Ratio	N/C_{SAP} Ratio	N^+/N_{tot} (%) Ratio	O–org/Me–O_x Ratio	C_{SAP}/Ti Ratio	SAP Thickness (nm)
pH 2	0.63	0.42	0.09	0.18	0.90	18.11	—	—	—
pH 4	0.57	0.37	0.05	0.14	0.77	15.57	—	—	—
pH 7	0.37	0.13	0.17	0.02	0.28	0.00	1.29	4.07	1.54
pH 10	0.56	0.37	0.06	0.16	0.87	10.99	—	—	—
pH 12	0.48	0.27	0.10	0.11	0.76	11.33	1.75	8.17	4.31

Calculated MUL thicknesses for pH 7 and pH 12 are reported in the last column of Table 2.

As for the semiquantitative analysis, C–N/C–C and N–C=O/C–N ratios are more or less constant in all samples, except for sample pH 7, where they are the lowest, as observed in Table 2. This could be due to peptide degradation or non-uniform deposition, as seen in microscopy images, since it has also a greater component related to oxidized carbons (COOH).

3.2.2. Analysis of SAP Monolayer (ML) Adsorbed at Different pH Values

As already observed for MULs, MLs samples (investigated by SR-induced XPS) show complex C1s signals due to the presence of different functional groups, as depicted in Figure 5b for the pH 2 sample. Similar to MULs, B.E. values were calibrated according to C–C aliphatic carbon (285.0 eV); all contributions show the expected value: main peak C–C 285.0 eV B.E., C–N ~ 286.3 eV BE, N–C=O (amide-like) ~ 288.3 eV BE and COOH ~ 289 eV BE. Atomic ratios, reported in Table 3, show a very high amount of aliphatic carbons with respect to MULs, probably due to the higher relative amount of contaminants with respect to the amount of immobilized peptide. However, the relative amounts of C1s spectral components are not influenced by the preparation pH, confirming that the molecular structure of the oligopeptides is always preserved in the anchoring process.

Table 3. Atomic ratios of the species present on ML. N^+: protonated nitrogen; Me–O_x: oxygen of metal oxides; O–org: organic oxygen.

ML Sample	C–N/C–C Ratio	N–C=O/C–C Ratio	COOH/C–C Ratio	N_{tot}/C_{tot} Ratio	N/C_{SAP} Ratio	N^+/N_{tot} (%) Ratio	O–org/Me–O_x Ratio	C_{SAP}/Ti Ratio	SAP Thickness (nm)
pH 2	0.44	0.25	0.10	0.11	0.77	13.47	24.76	2.91	1.22
pH 4	0.42	0.21	0.12	0.06	0.23	9.20	1.83	13.41	3.10
pH 7	0.34	0.12	0.10	0.04	0.58	4.13	1.04	1.76	1.28
pH 10	0.38	0.19	0.12	0.05	0.46	0.00	1.67	4.07	2.52
pH 12	0.32	0.16	0.11	0.04	0.42	0.00	0.96	2.15	1.32

As expected, all samples present a N1s signal; by applying a peak fitting procedure two main components related to protonated and unprotonated nitrogen atoms can be singled out, as already discussed for MULs. It is noteworthy that the intensity of protonated nitrogen decreases along with the increasing solution alkalinity, as expected (see Table 3). The N1s components trend is clearly observable in the following Figure 6.

Differently from MUL samples, O1s spectra show all the components observed for pristine samples, i.e., Me–O (~530 eV BE), C–O (~532 eV BE), and physisorbed water (~533 eV BE). The peptide coverage in MLs prepared at neutral and basic pH values (i.e., pH = 7, 10, 12), in fact, is not so thick as to completely screen the substrate signal, as observed in multilayers. This allows the layer thickness to be estimated from the substrate signal attenuation for ML prepared at these three pH values, as reported in Table 3. The trend observed in film thickness is supported by the rough titanium spectra, shown in the following Figure 7.

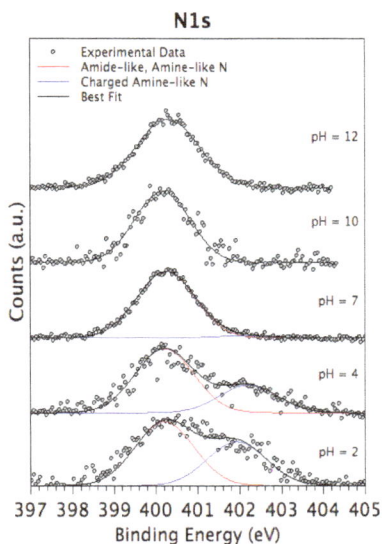

Figure 6. XPS N1s spectra collected on the four monolayer (ML) samples prepared with decreasing pH values.

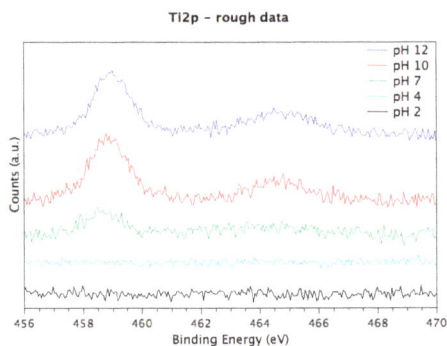

Figure 7. Rough Ti2p spectra, showing the attenuation of substrate signal with decreasing pH, indicative for a better peptide adsorption.

The observed experimental evidence (C1s components composition preserved in all samples, Ti2p signal attenuation increased with decreasing pH, film thicknesses increased at low pHs, protonated nitrogen component appearing at lower pHs) all concurs to confirm the oligopeptide stability upon anchoring at the titanium alloy surface, and to point out a better surface functionalization performance at low pH values, in good agreement with previous studies carried out on titania surfaces [50,51]. This effect could be related to the net charge of the EAbuK peptide; due to amidation of the carboxyl terminal, the investigated sample has a basic function more than an acidic one and a resulting net charge of +1, increasing peptide solubility at low pH.

3.2.3. Near Edge X-ray Absorption Fine Structure Spectroscopy

NEXAFS (Near Edge X-ray Absorption Fine Structure) is an X-ray absorption technique that, when performed in Angular Dependent mode (i.e., by measuring the X-rays absorption as a function of the incidence angle of the impinging photons with respect to the sample surface), can be usefully applied to evidence whether the peptide molecules are oriented with respect to the surface [51,67]. Data were collected at the C and N K-edges for both MULs and MLs. When the peptide is oriented and organized with respect to the surface a difference in radiation absorption is expected when the angle of the impinging radiation is changed from grazing to normal. The magic incidence is the peculiar angle at which the signal looks as if the peptide was not organized.

The carbon spectrum of MUL prepared at pH 2 is reported in the following Figure 8a; in C K-edge spectrum several structures can be observed in both π^* and σ^* resonances region. The sharp feature at about 288.7 eV is assigned to a C1s $\rightarrow$ π^* transition of C=O molecular orbital, the shoulder around 288 eV to a σ^* resonance by the C–H groups, additional features at $\approx$293 and $\approx$303 eV can be associated to 1 s $\rightarrow$ σ^* transitions by the C–C and respectively C=O molecular groups.

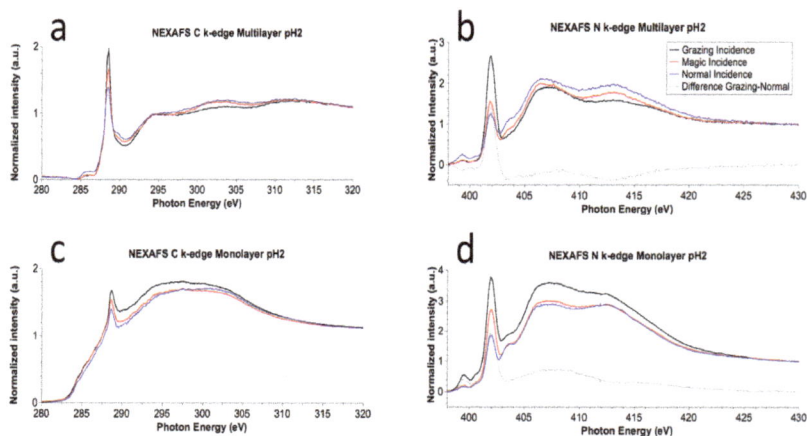

Figure 8. (**a**) C K-edge Near Edge X-ray Absorption Fine Structure (NEXAFS) spectrum of EAbuK multilayer deposited at pH = 2; (**b**) N K-edge NEXAFS spectrum of EAbuK at pH = 2; (**c**) C K-edge NEXAFS spectrum of EAbuK monolayer deposited at pH = 2; (**d**) N K-edge NEXAFS spectrum of EAbuK monolayer at pH = 2.

N K-edge spectra (Figure 8b, MUL pH 2, as an example) show the π^* (402 eV) and σ^* features (406 eV and ~413 eV) associated with the electronic transitions from N1s to the related antibonding molecular orbitals, as expected from the molecular structure and amino acids sequence [50,51].

The single transition $\pi \rightarrow \pi^*$ and the two $\sigma \rightarrow \sigma^*$ are represented by the sharp strong peak at 402 eV and 406eV and 412 eV respectively. It is noteworthy how the intensity of transitions $\pi \rightarrow \pi^*$ increases from normal to grazing incidence while the intensity of $\sigma \rightarrow \sigma^*$ transitions decreases.

ML spectra are completely analogous, as shown in Figure 8c,d for the ML prepared at pH = 2.

The angular dependence analysis evidences changes in the relative intensity for the π^* and σ^* at different angle resonances in all samples, both MUL and ML, a phenomenon that can be associated with molecular orientation with respect to the surface.

The dichroic behavior of the π^* band associated with the peptide bond allowed the tilt angle to be calculated between the π^* vector orbital of the peptide bond and normal to the surface, by using the equation reported by Stöhr [67] for threefold or higher symmetry substrates, with a polarization factor p = 0.95, and the intensity ratio $I_{20°}/I_{90°}$ determined for the selected resonance by peak fitting of the experimental data. For the samples prepared at pH 2, the calculated value for the angle gives rise to a value of the tilt angle between the peptide bond axis (the axis of the main chain) and the substrate surface of nearly $80°$ (ML $\theta = 80.2°$; MUL $\theta = 78.2°$) for a β-sheet conformation of the peptide backbone. Considering that the incertitude on the tilt angle evaluation is 15% of the calculated value, the two systems have approximately the same molecular orientation.

3.3. Microscopy Analysis

3.3.1. FESEM Analysis

FESEM analysis was performed on MULs in order to investigate whether the peptide aggregates give rise to any peculiar structure. The result is that the peptide covers evenly the surface without assembling in any microscopic or nanoscopic structure detectable with this technique. For all here reported images the yellow bar is 200 μm, except for Figure 9D where the bar is 5 μm. The pristine surface (Figure 9A) appears quite smooth and regular, and peptide immobilization at pH 2, 10 and 12 does not change the appearance (Figure 9B,F,G, respectively). At pH 7 (Figure 9E) the peptide multilayer does not evenly cover the surface, most likely making the layer appear thin at XPS analysis. This result is likely due to the presence of phosphate in solution, in fact the sample prepared without Hank solution is analogous to the other ones. The pH 4 sample (Figure 9C) shows the thickest layer: in fact the peptide surface is extensively cracked because of its thickness. The yellow square identifies the region magnified in Figure 9D, where it is evident how the peptide makes a really thick layer that detaches from the substrate.

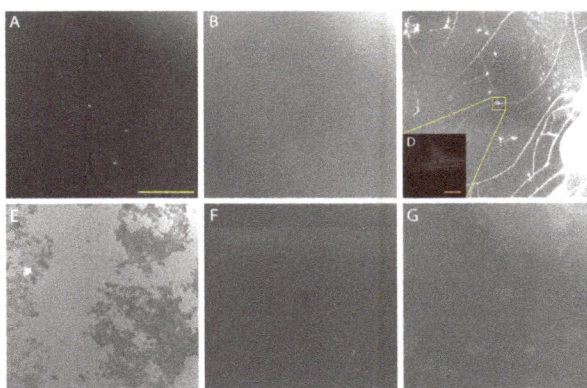

Figure 9. Field Emission Scanning Electron Microscopy (FESEM) images of MUL samples. (**A**): Pristine surface; (**B**): pH 2; (**C**): pH 4; (**D**): magnification of pH 4 region of interest; (**E**): pH 7; (**F**): pH 10; (**G**): pH 12. Yellow bar = 200 μm; Red bar (only D) = 5 μm.

3.3.2. AFM Measurements

Figure 10 shows AFM images acquired on all samples and on pristine surface. RMS values reported in Table 4 refer to the 900 μm^2 areas illustrated in Figure 10. It is evident how MUL samples show a more uniform surface with respect to the ML ones, especially at acidic pHs. According to the thickness evaluation reported in Tables 3 and 4, a thicker layer hides surface irregularity thus establishing an even surface. The pristine surface AFM image is shown in Supplementary Material Figure S2.

Monolayer   Multilayer

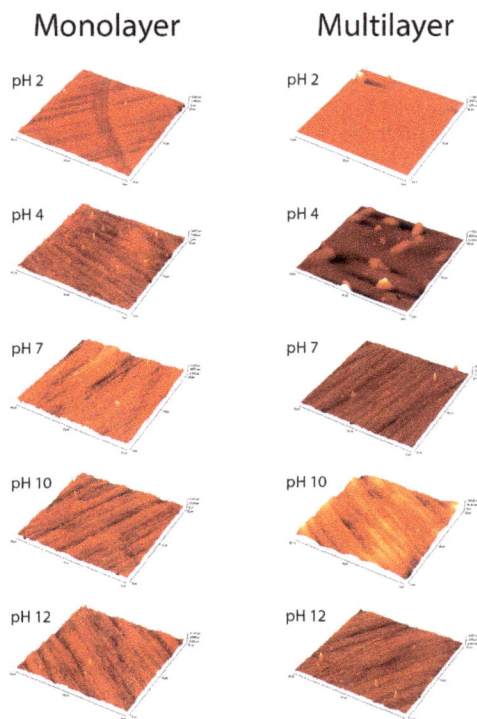

Figure 10. Atomic Force Microscopy (AFM) images of ML and MUL samples. It is evident how surfaces are more uniform at acidic pH. Pristine surface is shown in Supplementary Material Figure S2.

Table 4. Root mean square (RMS) roughness values for multilayer samples (MUL) and monolayer samples (ML).

Sample	RMS (μm)	
Clean Surface	0.0143	
	MUL RMS (μm)	ML RMS (μm)
pH 2	0.0181	0.0446
pH 4	0.0136	0.1067
pH 7	0.0155	0.0136
pH 10	0.0201	0.0141
pH 12	0.0127	0.0174

4. Conclusions

A new titanium alloy for orthopedic prostheses, containing Nb and Zr, was prepared by CCLM technique and characterized by various different techniques. Accurate characterization of the pristine alloy was carried out, revealing homogeneous mixing of the constituent elements (EDS), the presence of both α″ (orthorhombic) and β (disordered body-centered cubic) crystalline phases (XRD) and a TiO_2, Nb_2O_5, and ZrO_2 surface composition (XPS); moreover, electrochemical tests showed that Ti25Nb10Zr alloy corroded more heavily in Hank than in SBF solution, useful information for surface functionalization with peptides. We subsequently investigated the adsorption of EAbuK16, a self-assembling peptide, as a function of pH of the mother solution, both in the monolayer and multilayer. XPS analysis revealed that the amount of adsorbed peptide increases with decreasing pH of the investigated solutions. FESEM and AFM analysis confirmed the formation of a thick and homogenous overlayer at acidic pH. This effect is probably related to the net EAbuK charge of +1, which increases peptide solubility at low pH. NEXAFS investigation yielded evidence of molecular order and orientation of the peptide overlayer with respect to the substrate surface. Formation of an ordered SAP scaffold on the alloy surface should increase osteoblast adhesion to the material surface, thus improving osseo-integration.

Supplementary Materials: The following are available online at http://www.mdpi.com/2079-4991/8/3/148/s1, Figure S1: Total EDS mapping spectrum and relative distribution of the concentration of each constituent element on the Ti25Nb10Zr alloy surface performed in a one scanned zone on the surface, Figure S2: AFM image of Clean Surface Sample, Table S1: The elemental composition of the alloy determined by EDS measurements, Table S2: Binding Energies in eV of the atomic species present on the Clean Surface Sample and the relative atomic ratios, Table S3: Binding Energies in eV of the atomic species present on the MUL sample surface, Table S4: Binding Energies in eV of the atomic species present on the ML sample surface.

Acknowledgments: The authors acknowledge the CERIC-ERIC Consortium for access to experimental facilities and financial support. Part of the present work was supported under a grant of the Romanian National Authority for Scientific Research, CNCS—UEFISCDI, project number PN-II-PT-PCCA-2014-212 (OSSEOPROMOTE). We thank HZB for the allocation of synchrotron radiation beamtime.

Author Contributions: S.F., G.I. and C.B. conceived and designed the experiments; V.S. and M.S. performed the experiments and analyzed the data; M.D. and A.Z. synthesized the self assembling peptide, A.V. and M.B. prepared the alloy and investigated their phase composition, corrosion resistance and morphology after corrosion tests, and also the AFM of alloy pristine surfaces, T.S. helped and assisted the measurements at MSB, J.N. performed FESEM imaging.

Conflicts of Interest: The authors declare no conflict of interest. The founding sponsors had no role in the design of the study; in the collection, analyses, or interpretation of data; in the writing of the manuscript, and in the decision to publish the results.

References

1. Faria, A.C.L.; Rodrigues, R.C.S.; Rosa, A.L.; Ribeiro, R.F. Experimental titanium alloys for dental applications. *J. Prosthet. Dent.* **2014**, *112*, 1448–1460. [CrossRef] [PubMed]

2. Textor, M.; Sittig, C.; Frauchiger, V.; Tosatti, S.; Brunette, D.M. Properties and biological significance of natural oxide films on titanium and its alloys. In *Titanium in Medicine*; Springer: Berlin/Heidelberg, Germany, 2001; pp. 171–230.

3. Shah, F.A.; Trobos, M.; Thomsen, P.; Palmquist, A. Commercially pure titanium (cp-Ti) versus titanium alloy (Ti6Al4V) materials as bone anchored implants—Is one truly better than the other? *Mater. Sci. Eng. C* **2016**, *62*, 960–966. [CrossRef] [PubMed]

4. Dettin, M.; Zamuner, A.; Brun, P.; Castagliuolo, I.; Iucci, G.; Battocchio, C.; Messina, M.; Marletta, G. Covalent grafting of Ti surfaces with peptide hydrogel decorated with growth factors and self-assembling adhesive sequences. *J. Pept. Sci.* **2014**, *20*, 585–594. [CrossRef] [PubMed]

5. Liu, X.; Chu, P.K.; Ding, C. Surface modification of titanium, titanium alloys, and related materials for biomedical applications. *Mater. Sci. Eng. R* **2004**, *47*, 49–121. [CrossRef]

6. Özcan, M.; Hämmerle, C. Review: Titanium as a Reconstruction and Implant Material in Dentistry: Advantages and Pitfalls. *Materials* **2012**, *5*, 1528–1545. [CrossRef]

7. Brown, S.A.; Lemons, J.E. *Medical Applications of Titanium and Its Alloys: The Material and Biological Issues*; ASTM: West Conshohocken, PA, USA, 1996.

8. Elias, C.N.; Lima, J.H.C.; Valiev, R.; Meyers, A. Biomedical applications of titanium and its alloys. *JOM* **2008**, *60*, 46–49. [CrossRef]

9. Khorasani, A.M.; Goldberg, M.; Doeven, E.H.; Littlefair, G. Titanium in biomedical applications—Properties and fabrication: A review. *J. Biomater. Tissue Eng.* **2015**, *5*, 593–619. [CrossRef]

10. Veiga, C.; Davim, J.P.; Loureiro, A.J.R. Properties and applications of titanium alloys: A brief review. *Rev. Adv. Mater. Sci.* **2012**, *32*, 133–148.

11. Geetha, M.; Singh, A.K.; Asokamani, R.; Gogia, A.K. Ti based biomaterials, the ultimate choice for orthopaedic implants—A review. *Prog. Mater. Sci.* **2009**, *54*, 397–425. [CrossRef]

12. Sidambe, A.T. Biocompatibility of Advanced Manufactured Titanium Implants—A Review. *Materials* **2014**, *7*, 8168–8188. [CrossRef] [PubMed]

13. Katzer, A.; Hockertz, S.; Buchhorn, G.H.; Loehr, J.F. In vitro toxicity and mutagenicity of CoCrMo and Ti6Al wear particles. *Toxicology* **2003**, *190*, 145–154. [CrossRef]

14. Satoh, K.; Sato, S.; Wagatsuma, K. Formation mechanism of toxic-element-free oxide layer on Ti–6Al–4V alloy in d.c. glow discharge plasma with pure oxygen gas. *Surf. Coat. Technol.* **2016**, *302*, 82–87. [CrossRef]

15. Lecocq, M.; Félix, M.S.; Linares, J.-M.; Chaves-Jacob, J.; Decherchi, P.; Dousset, E. Titanium implant impairment and surrounding muscle cell death following neuro-myoelectrostimulation: An in vivo study. *J. Biomed. Mater. Res. Part B* **2015**, *103*, 1594–1601. [CrossRef] [PubMed]

16. Hussein, M.A.; Mohammed, A.S.; Al-Aqeeli, N. Wear Characteristics of Metallic Biomaterials: A Review. *Materials* **2015**, *8*, 2749–2768. [CrossRef]

17. Mahapatro, A.J. Metals for biomedical applications and devices. *J. Biomater. Tissue Eng.* **2012**, *2*, 259–268. [CrossRef]

18. Niinomi, M.; Nakai, M.; Hieda, J. Development of new metallic alloys for biomedical applications. *Acta Biomater.* **2012**, *8*, 3888–3903. [CrossRef] [PubMed]

19. Okazaki, Y.; Rao, S.; Asao, S.; Tateishi, T.; Katsuda, S.; Furuki, Y. Effects of Ti, Al and V concentrations on cell viability. *Mater. Trans. JIM* **1998**, *39*, 1053–1062. [CrossRef]

20. Zhang, L.-C.; Attar, H.; Calin, M.; Eckert, J. Review on manufacture by selective laser melting and properties of titanium based materials for biomedical applications. *J. Mater. Technol.* **2016**, *31*, 66–76. [CrossRef]

21. Li, Y.; Yang, C.; Zhao, H.; Qu, S.; Li, X.; Li, Y. New Developments of Ti-Based Alloys for Biomedical Applications. *Materials* **2014**, *7*, 1709–1800. [CrossRef] [PubMed]

22. Hao, Y.-L.; Li, S.-J.; Yang, R. Biomedical titanium alloys and their additive manufacturing. *Rare Met.* **2016**, *35*, 661–671. [CrossRef]

23. Wang, L.; Lu, W.; Qin, J.; Zhang, F.; Zhang, D. Influence of cold deformation on martensite transformation and mechanical properties of Ti–Nb–Ta–Zr alloy. *J. Alloys Compd.* **2009**, *469*, 512–518. [CrossRef]

24. Tane, M.; Hagihara, K.; Ueda, M.; Nakano, T.; Okuda, Y. Elastic-modulus enhancement during room-temperature aging and its suppression in metastable Ti–Nb–Based alloys with low body-centered cubic phase stability. *Acta Mater.* **2016**, *102*, 373–384. [CrossRef]

25. Stenlund, P.; Omar, O.; Brohede, U.; Norgren, S.; Norlindh, B.; Johansson, A.; Lausmaa, J.; Thomsen, P.; Palmquist, A. Bone response to a novel Ti–Ta–Nb–Zr alloy. *Acta Biomater.* **2015**, *20*, 165–175. [CrossRef] [PubMed]

26. Okazaki, Y. A New Ti–15Zr–4Nb–4Ta alloy for medical applications. *Curr. Opin. Solid State Mater. Sci.* **2001**, *5*, 45–53. [CrossRef]

27. Okazaki, Y.; Gotoh, E. Comparison of fatigue strengths of biocompatible Ti-15Zr-4Nb-4Ta alloy and other titanium materials. *Mater. Sci. Eng. C* **2011**, *31*, 325–333. [CrossRef]

28. Banerjee, R.; Nag, S.; Stechschulte, J.; Fraser, H.L. Strengthening mechanisms in Ti–Nb–Zr–Ta and Ti–Mo–Zr–Fe orthopaedic alloys. *Biomaterials* **2004**, *25*, 3413–3419. [CrossRef] [PubMed]

29. Nag, S.; Banerjee, R.; Fraser, H.L. Microstructural evolution and strengthening mechanisms in Ti–Nb–Zr–Ta, Ti–Mo–Zr–Fe and Ti–15Mo biocompatible alloys. *Mater. Sci. Eng. C* **2005**, *25*, 357–362. [CrossRef]

30. Liu, Y.Z.; Zu, X.T.; Qiu, S.Y.; Wang, L.; Ma, W.G.; Wei, C.F. Surface characterization and corrosion resistance of Ti–Al–Zr alloy by niobium ion implantation. *Nucl. Instrum. Methods Phys. Res. Sect. B* **2006**, *244*, 397–402. [CrossRef]

31. Chaves, J.M.; Florêncio, O.; Silva, P.S.; Marques, P.W.B.; Afonso, C.R.M. Influence of phase transformations on dynamical elastic modulus and anelasticity of beta Ti–Nb–Fe alloys for biomedical applications. *J. Mech. Behav. Biomed. Mater.* **2015**, *46*, 184–196. [CrossRef] [PubMed]

32. Miyazaki, S.; Kim, H.Y.; Hosoda, H. Development and characterization of Ni-free Ti-base shape memory and superelastic alloys. *Mater. Sci. Eng. A* **2006**, *438–440*, 18–24. [CrossRef]

33. Bai, Y.; Hao, Y.L.; Li, S.J.; Hao, Y.Q.; Yang, R.; Prima, F. Corrosion behavior of biomedical Ti–24Nb–4Zr–8Sn alloy in different simulated body solutions. *Mater. Sci. Eng. C* **2013**, *33*, 2159–2167. [CrossRef] [PubMed]

34. Fojt, J.; Joska, L.; Málek, J. Corrosion behaviour of porous Ti–39Nb alloy for biomedical applications. *Corros. Sci.* **2013**, *71*, 78–83. [CrossRef]

35. Okazaki, Y.; Rao, S.; Asao, S.; Tateishi, T. Effects of metallic concentrations other than Ti, Al and V on cell viability. *Mater. Trans. JIM* **1998**, *39*, 1070–1079. [CrossRef]

36. Li, Y.; Wong, C.; Xiong, J.; Hodgson, P.; Wen, C. Cytotoxicity of titanium and titanium alloying elements. *J. Dent. Res.* **2010**, *89*, 493–497. [CrossRef] [PubMed]

37. Matsuno, H.; Yokoyama, A.; Watari, F.; Uo, M.; Kawasaki, T. Biocompatibility and osteogenesis of refractory metal implants, titanium, hafnium, niobium, tantalum and rhenium. *Biomaterials* **2001**, *22*, 1253–1262. [CrossRef]

38. Hickman, J.W.; Gulbransen, E.A. Oxide films formed on titanium, zirconium, and their alloys with nickel, copper, and cobalt. *Anal. Chem.* **1948**, *20*, 158–165. [CrossRef]

39. Yilmazbayhan, A.; Motta, A.T.; Comstock, R.J.; Sabol, G.P.; Laid, B.; Cai, Z. Structure of zirconium alloy oxides formed in pure water studied with synchrotron radiation and optical microscopy: Relation to corrosion rate. *J. Nucl. Mater.* **2004**, *324*, 6–22. [CrossRef]

40. Motta, A.T.; Gomes da Silva, M.J.; Yilmazbayhan, A.; Comstock, R.J.; Cai, Z.; Lai, B. Microstructural characterization of oxides formed on model Zr alloys using synchrotron radiation. *J. ASTM Int.* **2008**, *5*, 1–20. [CrossRef]

41. Wang, B.L.; Zheng, Y.F.; Zhao, L.C. Electrochemical corrosion behavior of biomedical Ti–22Nb and Ti–22Nb–6Zr alloys in saline medium. *Mater. Corros.* **2009**, *60*, 788–794. [CrossRef]

42. Ishii, M.; Kaneko, M.; Oda, T. *Titanium and Its Alloys as Key Materials for Corrosion Protection Engineering*; Shin-Nittetsu Giho: Tokyo, Japan, 2002; pp. 49–56.

43. Abdel-Hady, M.; Fuwa, H.; Hinoshita, K.; Kimura, H.; Shinzato, Y.; Morinaga, M. Phase stability change with Zr content in β-type Ti–Nb alloys. *Scr. Mater.* **2007**, *57*, 1000–1003. [CrossRef]

44. Kim, J.I.; Kim, H.Y.; Inamura, T.; Hosoda, H.; Miyazaki, S. Shape memory characteristics of Ti–22Nb–(2–8)Zr(at %) biomedical alloys. *Mater. Sci. Eng. A* **2005**, *403*, 334–339. [CrossRef]

45. Wang, B.L.; Li, L.; Zheng, Y.F. In vitro cytotoxicity and hemocompatibility studies of Ti-Nb, Ti-Nb-Zr and Ti-Nb-Hf biomedical shape memory alloys. *Biomed. Mater.* **2010**, *5*, 44102. [CrossRef] [PubMed]

46. Kim, J.I.; Kim, H.Y.; Inamura, T.; Hosoda, H.; Miyazaki, S. Effect of annealing temperature on microstructure and shape memory characteristics of Ti–22Nb–6Zr (at %) biomedical alloy. *Mater. Trans.* **2006**, *47*, 505–512. [CrossRef]

47. Nayak, S.; Dey, T.; Naskar, D.; Kundu, S.C. The promotion of osseointegration of titanium surfaces by coating with silk protein sericin. *Biomaterials* **2013**, *34*, 2855–2864. [CrossRef] [PubMed]

48. Franchi, S.; Battocchio, C.; Galluzzi, M.; Navisse, E.; Zamuner, A.; Dettin, M.; Iucci, G. Self-assembling peptide hydrogels immobilized on silicon surfaces. *Mater. Sci. Eng. C* **2016**, *69*, 200–207. [CrossRef] [PubMed]

49. Gambaretto, R.; Tonin, L.; Di Bello, C.; Dettin, M. Self-assembling peptides: Correlation among sequence, secondary structure in solution and film formation. *Biopolymers* **2008**, *89*, 906–915. [CrossRef] [PubMed]

50. Iucci, G.; Battocchio, C.; Dettin, M.; Gambaretto, R.; Polzonetti, G. A NEXAFS and XPS study of the adsorption of self-assembling peptides on TiO$_2$: The influence of the side chains. *Surf. Interface Anal.* **2008**, *40*, 210–214. [CrossRef]

51. Battocchio, C.; Iucci, G.; Dettin, M.; Carravetta, V.; Monti, S.; Polzonetti, G. Self-assembling behaviour of self-complementary oligopeptides on biocompatible substrates. *Mater. Sci. Eng. B* **2010**, *169*, 36–42. [CrossRef]

52. Kokubo, T.; Takadama, H. How useful is SBF in predicting in vivo bone bioactivity? *Biomaterials* **2006**, *27*, 2907–2915. [CrossRef] [PubMed]

53. Gilewicz, A.; Chmielewska, P.; Murzynski, D.; Dobruchowska, E.; Warcholinski, B. Corrosion resistance of CrN and CrCN/CrN coatings deposited using cathodic arc evaporation in Ringer's and Hank's solutions. *Surf. Coat. Technol.* **2016**, *299*, 7–14. [CrossRef]

54. Mansfeld, F.; Oldham, K.B. A modification of the Stern—Geary linear polarization equation. *Corros. Sci.* **1971**, *11*, 787–796. [CrossRef]

55. Erika, G.; Ruslan, O.; Florian, S.; Hikmet, S.; Alexander, F. LowDosePES: An End-Station for Low-Dose, Angular-Resolved and Time-Resolved Photoelectron Spectroscopy at BESSY II. In Proceedings of the Scientific Opportunities with Electron Spectroscopy and RIXS, HZB/BESSY II, Berlin, Germany, 16–18 October 2017.

56. Moulder, J.F.; Stickle, W.F.; Sobol, P.E.; Bomben, K.D. *Handbook of X-ray Photoelectron Spectroscopy*; Physical Electronics Inc.: Eden Prairie, MN, USA, 1996.

57. Beamson, G.; Briggs, D. *High Resolution XPS of Organic Polymers, The Scienta ESCA300 Database*; John Wiley & Sons: Chichester, UK, 1992.

58. Briggs, D.; Seah, M.P. *Practical Surface Analysis, Vol. 1, Auger and X-ray Photoelectron Spectroscopy*; John Wiley & Sons: Chichester, UK, 1994.

59. Nannarone, S.; Borgatti, F.; De Luisa, A.; Doyle, B.P.; Gazzadi, G.C.; Gigli, A.; Finetti, P.; Mahne, N.; Pasquali, L.; Pedio, M.; et al. The BEAR Beamline at Elettra. *AIP Conf. Proc.* **2004**, *705*, 450–453.

60. Kim, H.Y.; Ikehara, Y.; Kim, J.I.; Hosoda, H.; Miyazaki, S. Martensitic transformation, shape memory effect and superelasticity of Ti–Nb binary alloys. *Acta Mater.* **2006**, *54*, 2419–2429. [CrossRef]

61. Bowen, A.W. Omega phase embrittlement in aged Ti-15%Mo. *Scr. Metall.* **1971**, *5*, 709–715. [CrossRef]

62. Sass, S.L. The ω phase in a Zr-25 at % Ti alloy. *Acta Metall.* **1969**, *17*, 813–820. [CrossRef]

63. Isaacs, H.S.; Ishikawa, Y. Current and Potential Transients during Localized Corrosion of Stainless Steel. *J. Electrochem. Soc.* **1985**, *132*, 1288–1293. [CrossRef]

64. Thirumalaikumarasamy, D.; Shanmugam, K.; Balasubramanian, V. Comparison of the corrosion behaviour of AZ31B magnesium alloy under immersion test and potentiodynamic polarization test in NaCl solution. *J. Magnes. Alloys* **2014**, *2*, 36–49. [CrossRef]

65. Song, G. Control of biodegradation of biocompatable magnesium alloys. *Corros. Sci.* **2007**, *49*, 1696–1701. [CrossRef]

66. Cotrut, C.M.; Parau, A.C.; Gherghilescu, A.I.; Titorencu, I.; Pana, I.; Cojocaru, D.V.; Pruna, V.; Constantin, L.; Dan, I.; Vranceanu, D.M.; et al. Mechanical, in vitro corrosion resistance and biological compatibility of casted and annealed Ti25Nb10Zr alloy. *Metals* **2017**, *7*, 86. [CrossRef]

67. Stohr, J. *NEXAFS Spectroscopy*; Gomer, C., Ed.; Springer Series in Surface Sciences; Springer: Berlin, Germany, 1991.

nanomaterials

MDPI

Article

Microscopic Views of Atomic and Molecular Oxygen Bonding with *epi* Ge(001)-2 × 1 Studied by High-Resolution Synchrotron Radiation Photoemission

Yi-Ting Cheng [1], Hsien-Wen Wan [1], Chiu-Ping Cheng [2,*], Jueinai Kwo [3,*], Minghwei Hong [1,*] and Tun-Wen Pi [4,*]

[1] Graduate Institute of Applied Physics and Department of Physics, National Taiwan University, Taipei 10617, Taiwan; ceo6120@gmail.com (Y.-T.C.); b00202022@ntu.edu.tw (H.-W.W.)
[2] Department of Electrophysics, National Chiayi University, Chiayi 60004, Taiwan
[3] Department of Physics, National Tsing Hua University, Hsinchu 30013, Taiwan
[4] National Synchrotron Radiation Research Center, Hsinchu 30076, Taiwan
* Correspondence: cpcheng@mail.ncyu.edu.tw (C.-P.C.); raynien@phys.nthu.edu.tw (J.K.); mhong@phys.ntu.edu.tw (M.H.); pi@nsrrc.org.tw (T.-W.P.); Tel.: +886-3-578-0281 (T.-W.P.)

Received: 21 February 2019; Accepted: 30 March 2019; Published: 4 April 2019

Abstract: In this paper, we investigate the embryonic stage of oxidation of an *epi* Ge(001)-2 × 1 by atomic oxygen and molecular O_2 via synchrotron radiation photoemission. The topmost buckled surface with the up- and down-dimer atoms, and the first subsurface layer behaves distinctly from the bulk by exhibiting surface core-level shifts in the Ge 3d core-level spectrum. The O_2 molecules become dissociated upon reaching the *epi* Ge(001)-2 × 1 surface. One of the O atoms removes the up-dimer atom and the other bonds with the underneath Ge atom in the subsurface layer. Atomic oxygen preferentially adsorbed on the *epi* Ge(001)-2 ×1 in between the up-dimer atoms and the underneath subsurface atoms, without affecting the down-dimer atoms. The electronic environment of the O-affiliated Ge up-dimer atoms becomes similar to that of the down-dimer atoms. They both exhibit an enrichment in charge, where the subsurface of the Ge layer is maintained in a charge-deficient state. The dipole moment that was originally generated in the buckled reconstruction no longer exists, thereby resulting in a decrease in the ionization potential. The down-dimer Ge atoms and the back-bonded subsurface atoms remain inert to atomic O and molecular O_2, which might account for the low reliability in the Ge-related metal-oxide-semiconductor (MOS) devices.

Keywords: Ge(001)-2 × 1; oxidation; synchrotron radiation photoemission

1. Introduction

Due to their high carrier mobilities, both the Ge and III-V compound semiconductors are channel materials that might replace silicon in p- and n-type metal-oxide-semiconductor field-effect transistors (MOSFETs) [1–10]. For the III-V metal-oxide-semiconductor (MOS), the established reports clearly show that a high-quality oxide/(In)GaAs interface leads to high-performance (In)GaAs MOSFETs in the drain currents and transconductances [9–13]. The premise of a high-quality interface requires a high-quality III-V surface (e.g., impurity-free and with a long-range order reflecting the nominal surface-atomic structure). Namely, the researchers are obligated to grow a high-κ dielectric oxide onto a well-defined reconstructed III-V surface. Years of researches on (In)GaAs(001) have concluded that chemical processes to the hetero-element surfaces would destroy the long-range order; the removal of the native oxides is out of the question for attaining a high-quality high-κ/III-V interface. To be specific, the as-grown (In)GaAs surface from a molecular-beam epitaxy (MBE) chamber or a metal organic chemical vapor deposition (MOCVD) reactor must be placed in direct contact with a dielectric oxide that was prepared with MBE or atomic layer deposition (ALD) without exposing it to air. For Ge, the surface pretreatment with assorted chemicals is typically done before dielectric deposition. The confidence in pretreatment is fully derived from the success of Si, where surface treatments are necessary in further device fabrication. Sun et al. have demonstrated that the Ge(001) surface becomes noticeably rough after exposure to aqueous HF and HCl solutions [14]. This raises concerns regarding the failure to advance Ge MOSFETs, which is presumably due to developers overlooking surface quality issues. Moreover, the engineering and surface-science communities, especially the synchrotron-radiation circle, have independently fought their wars, failing to maintain close communications with each other, thereby hindering efforts to achieve sufficient understanding of the interfaces.

The study of GeO_2/Ge becomes inevitable as a result of the desire to obtain an interface that is similar to that of SiO_2/Si with the oxide downsized to a few atomic layers. Researchers on Ge have been naturally inclined to take the established Si knowledge to propose the Ge MOS structure, and intensive research [15–20] and industry-scale development efforts in the Ge MOS have been undertaken to realize the product. Nevertheless, there is a dearth of fundamental research on Ge, largely because the Ge(001)-2 × 1 surface undergoes a similar reconstruction to that of Si(001)-2 × 1 and the surface and interfacial behaviors of Ge(001)-2 × 1 are believed to be no different from those of Si(001)-2 × 1. To our knowledge, this belief has never been questioned. In fact, the surface electronic structure of Ge(001)-2 × 1 reveals undeniable differences from that of Si(001)-2 × 1, where synchrotron radiation photoelectron spectroscopy (SRPES) shows the latter to exhibit distinct surface behaviors down to the third layer [21,22], but the former only to the second layer at room temperature (Figure 1) [23].

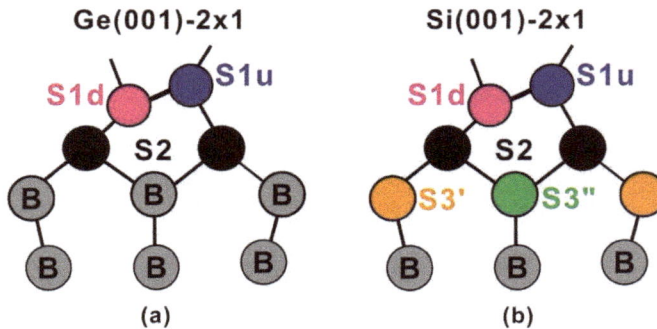

Figure 1. Schematic side-view drawings of (**a**) buckled Ge(001)-2 × 1 and (**b**) buckled Si(001)-2 × 1 surface. Symbols S1u, S1d, S2, and S3 stand for the up-dimer atom, down-dimer atom, atoms in the second surface layer, and atoms in the third surface layer, respectively. In Ge(001)-2 × 1, the S3 atoms show one electronic environment, but in Si(001)-2 × 1, they are differentiated by S3' and S3" electronic environments.

The lack of a detailed understanding regarding the Ge(001)-2 × 1 surface is an obstacle to understanding why Ge(001)-2 × 1 is unlikely to produce 1+ to 4+ charge states, as Si(001)-2 × 1 does [24]. The reported four Ge charge states were commonly observable by neutral beam oxidation [25], through chemically wetted treatments [26], during acid processes [27,28], or upon sudden exposure to a large amount of atomic oxygen on a 2 × 1 reconstructed surface [29]. It is odd to see that these thick Ge oxides respond differently with heat, with one showing an increase [30] and the other showing a decrease [27] in the strength of GeO_2 with increasing annealing temperatures. Nevertheless, the investigation of a thick Ge oxide film does not help us to fully understand the oxygen-contacted Ge-Ge dimers layer. Another method, an in vacuo process using supersonic molecular oxygen beams, only produces the 1+ and 2+ oxidation states in the sub-monolayer thickness [31]. The weak bonding of molecules on Ge(001) [32,33] and the confined adsorption site at the surface dimers without any insertion into the

Ge backbonds [34] have rendered the four charge states unlikely to be formed at the interfacial region of Ge(001).

O_2 is introduced into the MBE chamber to compensate for the loss of oxygen from bombarding the oxide target upon the deposition of a high-κ dielectric oxide onto a Ge(001) wafer,. The partial pressure of O_2 is unreactive to the (In)GaAs substrates [35], but it is not entirely unaffected at the Ge substrate; the ALD is operating at a high pressure and it generates similar concerns for the oxygen residual. The established records of initial O_2 adsorption on Ge(001) have made their comments based on the very first work by Fukuda and Ogina two decades ago, who suggested that one of the dissociated O atoms of O_2 sits at the bridge site and the other sits at a backbond of the Ge-Ge dimer [34,36]. However, the nondissociative O_2 adsorption has also been experimentally and theoretically proposed later in the literature [33,37,38]. Nevertheless, the established records fall short of experimental investigations on the interfacial electronic structure of Ge(001) with atomic O and molecular O_2, especially at the embryonic stage of adsorption. Worse yet, the investigations regarding an *epi* Ge(001)-2 × 1 surface are practically non-existent in the literature, the lack of which prevents us from improving the growth processes.

In this work, the *epi* Ge(001)-2 × 1 samples were exposed to atomic oxygen and high-purity O_2 at dosages from as small as 0.06 Langmuir (L) to as high as 400 L. SRPES probed the O/Ge system, which has been identified as a superb tool to probe the atom-to-atom interactions at the interface. The tunability of photon energies allows us to acquire the spectra with high surface sensitivity, which cannot be realized by conventional X-ray spectroscopy (XPS). Note that the present work does not deny the established XPS works, because, at its resolution, it is not possible to trace a surface component with a surface core-level shift (SCLS) as small as 150 meV. At a low concentration, the SRPES can provide the response of each atom in a buckled dimer to oxygen. For O_2/Ge(001), we found that the O_2 molecules become dissociated upon reaching the Ge surface. One of them would remove the up-dimer atom from the surface and the other bonds with the Ge atom in the subsurface layer. Two oxygen-induced chemical states are then resolved in the Ge 3d and O 1s core-level spectra. For atomic oxygen on Ge(001), only the charge-enriched up-dimer atoms will accept O into the vicinity, whereas the charge-deficient down-dimer atoms remain inert to O. Furthermore, the up-dimer atom donates its excess charge to the bonded oxygen atom under the intact dimer bond. Consequently, the charge environment becomes similar to that of the down-dimer atom. The present experimental results provide direct evidence that the unpassivated down-dimer atoms might account for the reliability issue that is related to Ge MOS devices [39,40].

2. Materials and Methods

2.1. Sample Preparations

The MBE technique was used to grow *epi* Ge(001)-2 × 1 layers on a Ge substrate. The Sb-doped n-type Ge(001) wafer with a resistivity of 0.31–0.34 Ω-cm was dipped in 2% diluted HF solution and then rinsed in de-ionized water. The wafer was then immediately loaded into a multi-chamber MBE/analysis system [41], in which it was annealed in a UHV at 600°C for 20 min to achieve an ordered surface. The quality of the sample was assessed based upon the streaky reconstructed reflection high-energy electron diffraction (RHEED) patterns [42]. Afterwards, 7-nm thick Ge was grown on the annealed Ge substrate while using an effusion cell in an MBE chamber at a UHV pressure of less than of 2×10^{-10} Torr. Using this approach, the Ge epi-layer shows a greatly improved 2× RHEED pattern with streaky diffraction spots and more distinct Kikuchi arcs. The sharper and more intense diffraction patterns indicate the attainment of a morphologically flatter and more atomically ordered surface [2]. The samples were stored in vacuo in a portable UHV module to transport to the nearby National Synchrotron Radiation Research Center for SRPES measurements. The photoelectrons were collected with a 150-mm hemispherical analyzer (SPECS, GmbH) in a μ-metal chamber with a base pressure that is better than 1.2×10^{-10} Torr. The overall instrumental resolution was greater than 60 meV. Silver film freshly grown from an electron gun before the measurements was used to determine the energy reference. Atomic oxygen was generated through a commercial cracker (SPECS, GmbH), and length of time on cracking and the chamber pressure determined the dosage (L).

2.2. Data Analysis

The objective of SRPES experiments on semiconductor surfaces is to relate the observable features with the known properties of the reconstructed surfaces. A Voigt function line commonly represents the photoemission component; that is, it is a convolute of the Lorentzian and Gaussian functions. A model function that is assumed to consist of a multiplicity of overlapping components should be set up and fit accordingly to the data by the least squares method. Constraints are necessary to reduce the ambiguity of a fit, such as the lifetime width, spin-orbit splitting, and ratio essentially identical for all of the components. The three parameters are, for each spin-orbit pair, its position, height, and Gaussian width. The amplitude of the line is, in principle, related to the areal density and the location of the atoms within the surface layer through the inelastic mean-free path (IMFP). Since the areas of Voigt function lines are not proportional to the product of peak height and Gaussian width, it will be necessary to numerically integrate the area to set the peak amplitude. In the buckled dimer reconstruction of the Ge(100) surface, each layer contains an equal number of atoms. Those that are in the surface layer exist in two states in equal number. In Si(001), there are features that divide the atoms in both the first and second subsurface layers into two classes (Figure 1). However, the atoms in the second subsurface layer have been considered to be identical in Ge(001), which is in contrast to the case in Si(001). In other words, only the first two top surface layers of Ge(001) are regarded as behaving differently from the bulk.

The algorithm that has been successfully set up to analyze the complicated line shape of the Si 2p core-level spectrum will be mainly used here to analyze the Ge 3d core-level spectra. The details can be found elsewhere [21,22]. In brief, the assumption of layerwise attenuation through the parameters of the escape depth (λ) and the layer spacing (d) correlated by $x = \exp(-d/\lambda)$ gives the fractional areal strengths of the first and second layers to be $1 - x$ and $x \times (1 - x)$, respectively. Given that the λ value is approximately 4 Å at a photon energy of 80 eV, the relative areal intensities, in the fit of the first layer (S1), the second layer (S2), and the bulk (B) in the normal-emission spectrum are expected to be 28%, 24%, and 48%, respectively. Due to the layerwise attenuation models only estimating the maximum fractional intensity of the second layer as 0.24, an unconstrained model function that gives a larger fractional intensity for the second layer can be immediately discarded.

3. Results and Discussion

3.1. Clean epi Ge(001)-2 × 1 Surface

The surface atoms naturally exhibit electronic structures that are distinct from the bulk atoms. Core-level photoemission is highly sensitive to the surface electronic structure, which is manifested by a SCLS from the bulk. If the surface line lies in a lower binding energy than the bulk, the shift is conventionally regarded as a negative shift, if higher, then the shift is said to be positive. In the specific Ge studies, efforts have been made to interpret the shallow Ge 3d core-level spectrum [23,43–47]. The lack of consensus in the SCLSs of the dimerized surface atoms in the established reports is rather unusual, despite the similar line shapes in these room-temperature data and fewer contributed line components in the Ge 3d state than those in the Si 2p state [22,44]. Some authors have suggested that the dimers are unbuckled, thereby giving rise to only one S1 component in the spectra [43]. The proposals for a dimerized surface differ; the shifted signs of the down atoms have been reported as positive [43–45] and as no shift at all [46,47]. It is not only the surface component(s) that have been reported differently by various research groups; there is also disagreement between groups regarding the line position of the second surface layer, which has been reported as having both a positive [23] or negative shift [43–47]. As follows, we show a resolution by approaching the line shape of an *epi*

Ge(001)-2 × 1 surface assisted with the physical IMFP effect. Note that the existing reports that are cited above dealt with chemically treated surfaces.

Figure 2a displays SRPES-acquired *epi* Ge 3d core-level spectra in normal and 65° off-normal emissions at room temperature. It is worth noting that the data presented in Figure 2 were collected after the samples had been in the portable UHV module for 48 h. The absence of an oxidation state suggests that the *epi* Ge(001) surface is rather stable in vacuum. If a similar arrangement is applied to Si(001), then the surface dangling bonds will certainly be oxidized.

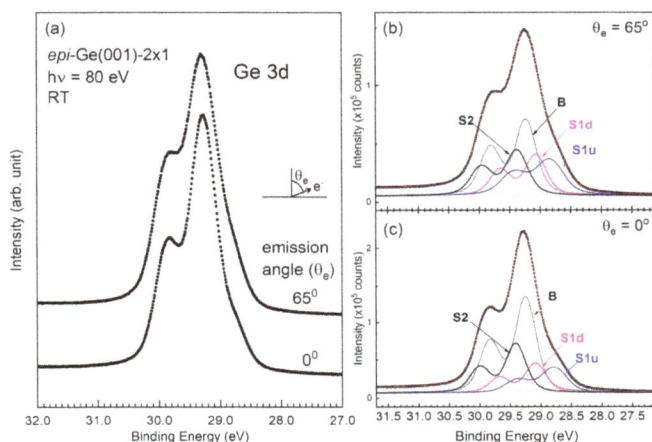

Figure 2. (**a**) Photoemission data from as-grown *epi* Ge(001)-2 × 1 taken in normal ($\theta_e = 0°$) and off-normal ($\theta_e = 65°$) emission at photon energies (hν) = 80 eV at room temperature. (**b**) A simultaneous fit to the 65°-emission spectrum. (**c**) A simultaneous fit to the normal-emission spectrum.

As shown in Figure 2a, a component lies clearly on the low-energy shoulder of the bulk component. Moreover, a change in line shape with emission angles is also observed, which is in contrast to the literature, where is it reported that little change in the spectral line shape occurs at different energies and emission angles [23,44–46]. Namely, the valley region increases in strength with increasing emission angles. Clearly, the *epi* Ge(001) surface shows a distinct surface electronic structure by SRPES that cannot be overlooked upon dielectric deposition. Note that the acquired XPS spectrum becomes broadened due to poor energy resolution, and the IMFP effect renders the contribution of surface emission rather small [42].

A fit is necessary to extract the embedded components of the *epi* Ge 3d line shape. A preliminary analysis that uses a proper model function suggests the need for four Voigt function-like components, which reside on a parameterized background and separate energy-loss tail. Four parameters represent the background: a constant, a slope, and a two-parameter power-law. This background function successfully represents the attenuated electron from a shallow Ge 3d level, where the loss tail is due to the electron-hole pair excitations in the spectral energy range. The presence of such a tail does not affect the primary line structure, because the onset of electron-hole pair excitations commences at 1.2 eV away from the bulk line. Note that the background function must be incorporated in a fit. Subtraction of the background before a fit should be avoided as it ignores the electron-hole pair excitations.

Figure 2b,c for $\theta_e = 65°$ and 0°, respectively, plot a representative fit to the room-temperature data, respectively. In this fit, the spin-orbit splitting is 0.593 and the spin-orbit branching ratio varies greatly from 0.514 in normal emission to 0.626 in $\theta_e = 65°$. The lifetime width is 0.172 eV. The low branching ratio has been explained as being due to the photon energy bearing near the photoemission threshold [48,49]. The binding energy of the bulk Ge 3d state is resolved at 29.25 eV. The SCLSs of the S1u, S1d, and S2 components are −0.462, −0.159, and +0.165 eV in the normal emissions, respectively.

The S2 component solely originates from the first subsurface layer [21,23]. The resultant intensities of the three surface components follow the expectation from the IMFP effect.

Decades of researches examining Si(001)-2 × 1 and Ge(001)-2 × 1 came to the consensus that the final-state theory explains the Si 2p and Ge 3d core-level spectra better than the initial-state theory. The final-state picture includes a crystal ensemble with a created core hole and a photo-excited electron. Pehlke and Scheffler [50] calculated the final-state values for the S1u and S1d components to be −0.67 and −0.39 eV away from the bulk, respectively. For one's reference, the initial-state values for the S1u and S1d components were −0.50 and +0.27 eV, respectively. A large screening shift was found for the 3d core electrons that were emitted from the down-dimer atoms. This was hypothesized to be the unoccupied dangling-bond state coming from the down-dimer atoms that were pulled down due to the influence of the core hole. The down-shifted dangling-bond state becomes populated by electrons from the Fermi level, thereby giving rise to effective screening. The lower value that was found in the present work suggests imperfect screening in the physical process [50]. Note that the phenomenon is less apparent in the 3d core electrons from the up-dimer atoms, because the corresponding dangling-bond state is already occupied. The final-state model predicts that the SCLS of the S2 component has a negative line position [50,51], which disagrees with the results that are presented in Figure 2a.

3.2. Adsorption of Molecular Oxygen on epi Ge(001)-2 × 1 Surface

Figure 3a shows the Ge 3d core-level spectra with various dosages of O_2 on the *epi* Ge(001)-2 × 1 surface at room temperature, being taken with 80-eV photon energy in normal emission. The inset plots the O 1s core-level spectra with selective O_2 dosages, where two O 1s states, O(I) and O(II), are resolved 1 eV apart. As displayed in Figure 3a, the adsorption of O_2 alters (not drastically) the line shape of the Ge 3d state, and the development of a germanium-oxide state on the high binding energy side of the bulk state is not as significant as that of SiO_x in Si(001)-2 × 1. The latter shows equally spaced peaks that gradually increase in intensity with increasing O_2 dosage [24]. The molecular oxygen immediately affects the S1u component in Figure 3, suggesting that the oxygen makes direct contact with the up-dimer atoms. As can be seen in Figure 3a, the emission from the S1u state becomes broadened but it remains observable at great coverages. Attempting to have O_2 covering all of the surface-dimerized atoms is unlikely, even 1000 L of O_2 exposure is of little help here [52]. In other words, the O_2/Ge(001)-2 × 1 interface is unlikely to result in high-oxidation states. In comparison, just 10 L of O_2 on Si(001)-2 × 1 have already developed four oxidation states with the highest one at the surface region and the smallest one on the bottom of the oxide layer [24]. Note that the surface Si-Si dimers remain detectable at 10 L of O_2 dosage, which suggests the local distribution of the SiO_x.

The affected Ge atoms would either appear as an induced component in the spectra or be removed from the surface. The induced Ge-O component would show increased binding energy below the bulk component, which is embedded in the spectral line envelope. Other than the noticeable feature lying an eV from the bulk line, another component that sits in the valley region is realized if the spectra of Figure 3a are normalized to the bulk line. Hence, six components are needed to represent the Ge 3d core-level spectra in the model function, four for the O_2-free, and two for the O_2-affected atoms. The number of O_2-induced components is consistent with two O 1s states, as shown in the inset of Figure 3a.

Further analysis of the line shape with a fit would allow us to determine whether the contacted oxygen is in an atomic or molecular form. Figure 3b,c display a fit to the 80 L and 2 L curves of Figure 3a, respectively. The induced GeO(I) and GeO(II) components are, respectively, resolved at approximately +0.355 and +1.333 eV, which gradually increase in intensity with increasing O_2 dosages. One might question the necessity of the GeO(I) component in Figure 3c, because its intensity is rather small. If the GeO(I) component was excluded in the model function, the S1d component would end up with an unphysical enhancement in intensity (~20%). The appearance of the GeO(I) component in the embryonic stage of O_2 adsorption suggests the immediate attachment of O_2 onto the up-atom dimers upon arriving at the Ge(001) surface. That is to say that no mediate stage exists in the O_2 adsorption on

Ge(001) [53]. In a fit, the S2, as well as the S1u component, shows a gradual decrease in intensity with O_2 dosages.

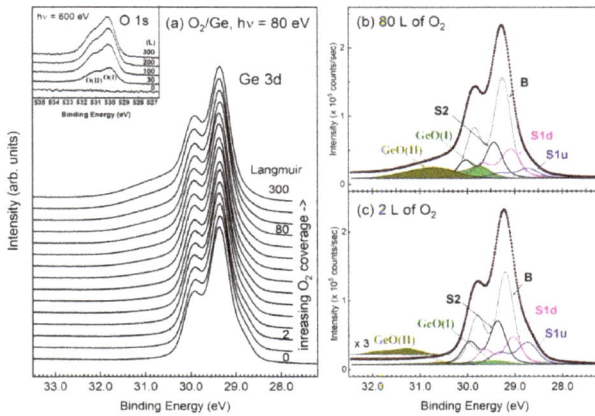

Figure 3. (**a**) The Ge 3d core-level spectra with various dosages of O_2 on an *epi* Ge(001)-2 × 1 surface at room temperature; (**b**) a fit to the Ge 3d core-level spectrum with 80 L of O_2 on Ge(001)-2 × 1; and, (**c**) a fit to the Ge 3d core-level spectrum with 2L of O_2 on Ge(001)-2 × 1. (Panels (**a**) and (**c**) reprinted with permission from [42], Copyright The Japan Society of Applied Physics, 2018.)

In a separate experiment, the deposition of a high-κ dielectric oxide onto the present 300-L O_2/Ge(001) surface has reduced the GeO(II) component (data not shown), which has a behavior that is similar to that of GaAs(001). That is to say, the GeO(II) is not an interfacial O/Ge component. The phenomenon indicates that the Ge atoms in GeO(II) are those that originated at the S1u up-dimer atoms now being set free to become the GeO(II) film. Once an S1u atom is freed, the other dissociated O atom immediately fills the vacancy and then bonds with the S2 atom underneath, thereby giving rise to the GeO(I) component. As shown in Figure 3b,c, the S1d component remains virtually in the same areal intensity as in the clean surface, which suggests that it does not follow the reacted pathway of the S1u component. Figure 4a presents the schematic drawings that summarize the interaction of O_2 with an up-dimer atom. The drawing is self-explanatory; the incoming molecular oxygen would disrupt the Ge-Ge dimer, so that one of the oxygen atoms would remove the up-dimer atom and the other immediately bonds with the nearby underneath subsurface Ge atom.

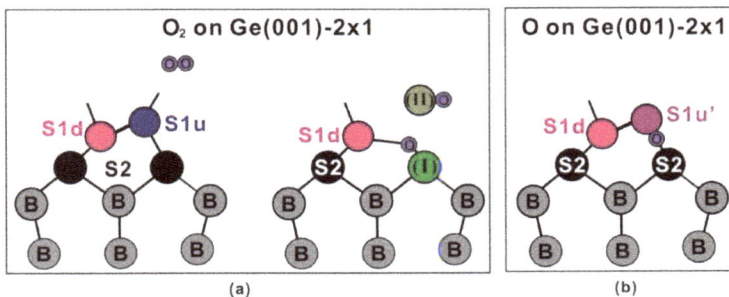

Figure 4. Schematic drawings of (**a**) O_2 onto Ge(001)-2 × 1, and (**b**) atomic O on Ge(001)-2 × 1 bonding configuration.

The preferential inclination of the O_2 molecules onto the up-dimer Ge atoms is certainly not found in the silicon counterpart and the previous reports of the chemically treated Ge surfaces [33,34,36,37]. On the one hand, the non-dissociative model should be forsaken, because it would result in only one GeO and one O 1s component. The O_2-dissociative model, on the other hand, predicted a non-destructive effect on the Ge-Ge dimers, with one oxygen atom sitting at the top bridge site and the other the backbond of the surface dimer. This assignment has the fundamental difficulty that the bridged O would revert the buckled dimer to a symmetric configuration, and a change in the work function should consequently be observed due to the change of the dipole within the dimerized atoms. Our separate measurements of the valence-band spectra and the cutoff region show gradual increases of the O 2s state and an unshifted zero kinetic-energy point with various O_2 exposures, respectively. The latter indicates that the work function of the O_2/Ge interface stays at the same value as that of the clean surface.

3.3. Adsorption of Atomic Oxygen on epi Ge(001)-2 × 1 Surface

Figure 5 displays a series of atomic-O covered *epi* Ge 3d core-level spectra that were taken with 80 eV photons at normal emission at room temperature. The dense plots in the bottom part of Figure 5 are the result of finely incremental O dosages below 1.86 L, and the upper part of the figure shows the curves over a dosage of 50 L. As can be seen in Figure 5, a gradual decrease in the surface intensity of the S1u state with increasing atomic O coverages suggests an immediate affiliation of the oxygen atoms with the surface dimers. As can be seen in Figure 5, the 400-L spectrum is not significantly different from the 1.86-L spectrum, suggesting that the growth of atomic O on *epi* Ge(001)-2 × 1 is a self-limiting process. Furthermore, the Ge 3d spectra display a peak shift of 40 meV towards a lower binding energy upon early appearance of oxygen atoms, which is later compensated for when the surface dimers have adsorbed the maximal amount of atomic O.

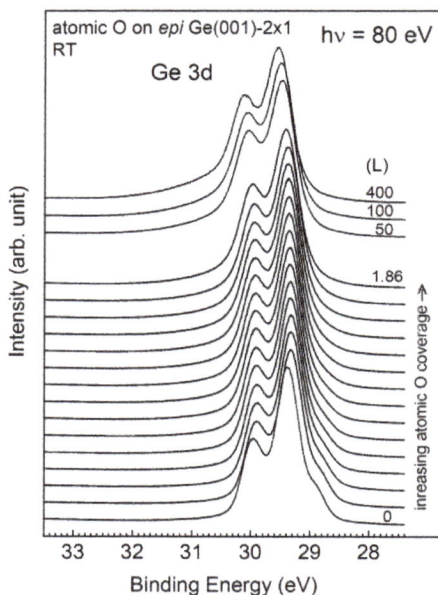

Figure 5. Development of the Ge 3d core-level spectra with various dosages of atomic O on an *epi* Ge(001)-2 × 1 surface at room temperature.

Figure 6 displays a fit for the representative curves reflected in Figure 5. Upon the early appearance of atomic O on the *epi* Ge(001)-2 × 1 surface, the line shape shows no great difference from that of the clean surface. Hence, we employ the model function of the clean surface to analyze the O-affected line spectra with great success. In contrast to the case of O_2 on Ge(001)-2 × 1 above, the O-bonded Ge atoms do not occur in an energy position that is lower than the bulk atoms, but they appear on the two dimer-related components. Note that the increased binding energy under expectation should be referred to the S1u component, not the bulk component. As can be seen in Figure 6, the (S1u'+S1d) component gradually increases in strength at the expense of the S1u component. Moreover, the S2 component remains virtually unaltered in intensity, even after the 400 L exposure to atomic oxygen (data not shown). The O atoms confine the reaction to the topmost surface layer and even restrict themselves to the top of the up-dimer atoms (see later). The oxidation path of atomic O on Ge(001)-2 × 1 certainly develops on its own, and it does not behave similarly to that of O_2 on either Si(001)-2 × 1 or Ge(001)-2 × 1. Oxygen is readily available to accept a full charge from Si to allow the interface to form a SiO_x film with x = 1 to 4 [24]. However, only a partial charge of the Ge surface atoms is needed to transport to the adsorbed oxygen atoms.

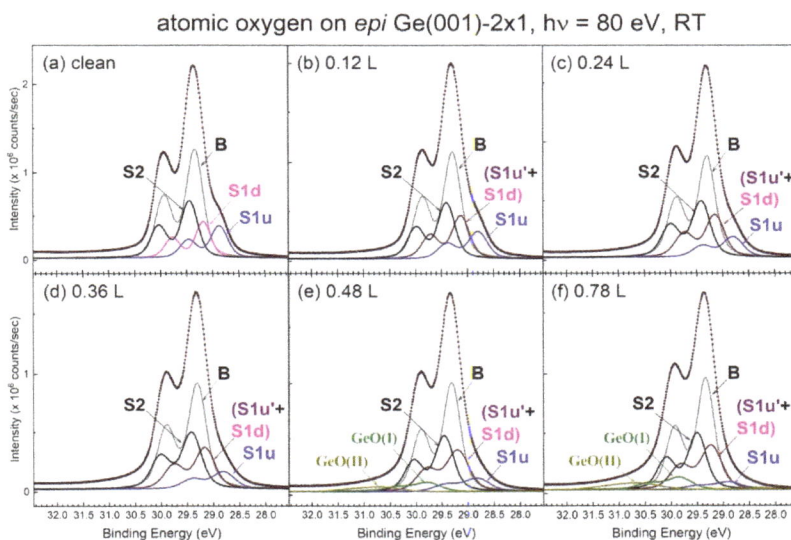

Figure 6. A fit to the representative Ge 3d core core-level spectra in Figure 5 emphasizing the initially low atomic-O coverages.

As the (S1u'+ S1d) symbol suggests, it consists of two states, the S1d atoms and the other from the O-bonded S1u atoms (S1u'), which makes the component 50% wider (e.g., 0.48 L) than the S1d component on the clean surface. The O-bonded S1u component appears to be fairly close to the S1d component in terms of energy position, thereby doubling the line intensity in the region. Attempts to split the S' component into two components yielded too small of an intensity in the induced S1u component. Direct evidence to support this speculation is derived from the information that is provided by the surface dipole. As is known, the buckled dimer results in a charge transfer from the down-dimer atom to the up-dimer atom, thereby giving rise to a dipole moment that was oriented with the positive end inwards in the asymmetric dimer. This will lead to an increase in the ionization potential (IP) of the buckled surface when compared with the unbuckled Ge(001)-2 × 1 surface. The change in IP upon O adsorption suggests an induced dipole. Due to the atomic oxygen accepting a charge from the up-dimer atom, the positive end of the O-Ge dipole moment orients the Ge atom. As a result, the IP becomes large in magnitude when the O atom resides on the up-dimer atom. The IP in a semiconductor is determined through the measurement of the spectral width (W) at a given photon energy (hν). The former W is the energy separation of the valence band maximum (VBM) and the cutoff of the photo-ejected electrons. The IP value is directly determined without any assumption by subtracting hν from W [54,55].

Figure 7 plots the spectral development of the cutoff and valence band regions. As shown in Figure 7, the onset of the cutoff gradually moves towards the lower kinetic energies with increasing atomic O dosages. The extrapolation of the rising slope with the constant background line pins down the cutoff position, and the IP development of each dosage with the nearly fixed valence band maximum (VBM), is plotted in the inset of Figure 7. The IP value decreases from 5.31 eV in the clean surface to 5.02 eV after 0.78 L of atomic O dosage. The result of the decreased IP is contrary to the expectation that the O atom is above the Ge atom. Note that the final IP occurs when the atomic O bonds with the up-dimer atoms fully. Although the binding energies of components S1u' and S1d are close to each other, the O-bonded dimer is still manifested in the buckled orientation. The placement of the atomic O at the backbond site is in accordance with the upward dipole direction that is found in the present study, as well as the theoretical calculation [56]. In the latter, it has been claimed that the highest energy-occupied surface states were exclusively backbond states [56]. Figure 4b shows the schematic drawing of atomic O/Ge(001)-2 × 1 bonding configuration. As shown in Figure 6e,f, the GeO(I) and GeO(II) components begin to appear at the Ge(001) surface, meaning that the O_2 has come into play on adsorption. This is because the present setup for breaking the molecular oxygen into its atomic form is not 100% efficient, according to the mass spectrum from a residual gas analyzer. However, the small O_2 residual does not affect the IP of the system, as mentioned above.

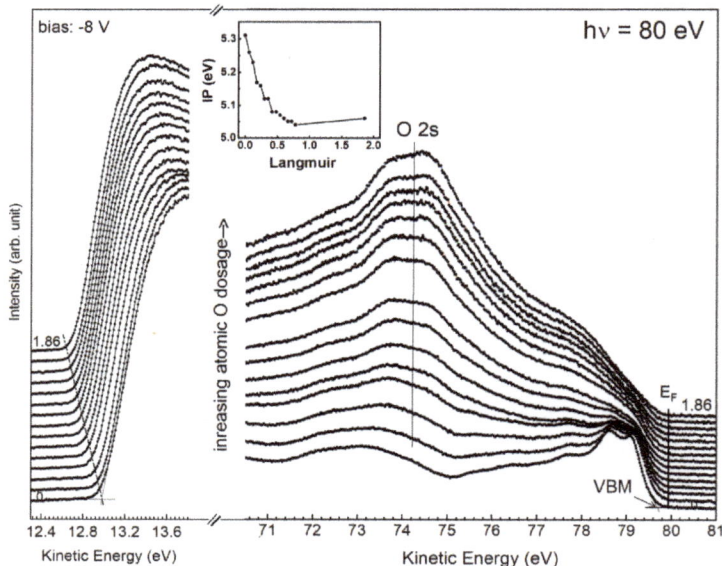

Figure 7. The acquired valence-band and cutoff spectra taken with 80 eV photon energy with various atomic-O coverages. The inset shows the change of the ionization potential (IP).

As can be seen in Figure 6, both the S1d and S2 components maintain the original energy positions with the unchanged SCLS signs in the atomic O-covered spectra. This suggests that the topmost dimers layer serves as a charge-enriched layer, while the subsurface layer serves as a charge-deficient layer. This indicates that a charge redistribution occurs between the first two surface layers. If this is indeed the case, then the Ge 3d core-level spectra faithfully show that the sign of the shift is negative in the first surface layer and positive in the second surface layer.

4. Conclusions

An early stage of oxidation of an *epi* Ge(001)-2 × 1 surface by atomic as well as molecular oxygen is presented while using high-resolution synchrotron radiation photoemission as a probe at room temperature. The pristine reconstructed Ge(001)-2 × 1 surface was first presented by setting up a proper model function to analyze the Ge 3d core-level spectrum, which is necessary for the later observation of oxidation development. In fact, only the first two top surface layers are considered to behave distinctly from the bulk. However, the topmost surface is buckled with one atom moving upward and the other downward. A charge transfer indeed occurs in between with the up-dimer atom being in a charge-enriched state, and the down-dimer atom a charge-deficient state. The charge environment plays a significant role for the oxygen atoms to preferentially react with the up-dimer atoms. For O_2, it is immediately dissociated without a mediated stage, and it simultaneously causes the up-dimer atom to exit the dimerized state. A dissociated oxygen bonds with the freed Ge atom, and the other inclines to be positioned at the site of the freed up-dimer atom, and bonds with the underneath Ge atom in the subsurface layer. The down-dimer atoms and those in the subsurface layer are inert to O_2. For atomic O, it selectively sits at the position between the up-dimer atom and the atom in the subsurface layer without causing any bond to be broken. Similar to the O_2 case, the down-dimer atoms are resistant to the effect of atomic oxygen. The reconstructed 2 × 1 configuration remains intact upon atomic O adsorption. Unlike the O/Si case, the adsorbed oxygen atoms accept a partial charge from the contacted Ge atoms. The full coverage of the up-dimers atoms with oxygen renders the surface layer into a uniform electronic state with excess charge. The second surface layer stays in the electronic state with deficient charge. The experimental results presented here might explain some of the reliability problems that are associated with the Ge MOS devices.

Author Contributions: M.H. and J.K. supervised the project. Y.-T.C. acquired the SRPES data, and plotted the valence band spectra. H.-W.W. grew the *epi* Ge(001)-2 × 1 samples. T.-W.P. did the fit and wrote the manuscript. All the authors revised the manuscript with eminent inputs from C.-P.C. on the interpretation of the ionization potential. All the authors approved the final version of the manuscript.

Funding: This work was supported by MOST 105-2112-M-213-007-MY3, 105-2112-M-007-014-MY3, 107-2119-M-002-049-, 107-2923-M-002-003-, and 107-2622-8-002-018 of the Ministry of Science and Technology in Taiwan.

Conflicts of Interest: The authors declare no conflict of interest.

References

1. Chu, L.K.; Chu, R.L.; Lin, T.D.; Lee, W.C.; Lin, C.A.; Huang, M.L.; Lee, Y.J.; Kwo, J.; Hong, M. Effective passivation and high-performance metal-oxide-semiconductor devices using ultra-high-vacuum deposited high-k dielectrics on Ge without interfacial layers. *Solid-State Electron.* **2010**, *54*, 965. [CrossRef]
2. Chu, R.L.; Liu, Y.C.; Lee, W.C.; Lin, T.D.; Huang, M.L.; Pi, T.W.; Kwo, J.; Hong, M. Greatly improved interfacial passivation of *in-situ* high k dielectric deposition on freshly grown molecule beam epitaxy Ge epitaxial layer on Ge(100). *Appl. Phys. Lett.* **2014**, *104*, 202102.
3. Dimoulas, A.; Tsoutsou, D.; Galata, S.; Panayiotatos, Y.; Mavrou, G.; Golias, E. Ge surfaces and its passivation by rare earth lanthanum germanate dielectric. *ECS Trans.* **2010**, *33*, 433–446.
4. Hu, S.; McDaniel, M.D.; Posadas, A.; Hu, C.; Wu, H.; Yu, E.T.; Smith, D.J.; Demkov, A.A.; Ekerdt, J.G. Monolithic integration of perovskites on Ge(001) by atomic layer deposition: A case study with SrHf$_x$Ti$_{1-x}$O$_3$. *MRS Commun.* **2016**, *6*, 125–132. [CrossRef]
5. Lu, C.; Lee, C.H.; Zhang, W.; Nishimura, T.; Nagashio, K.; Toriumi, A. Structural and thermodynamic consideration of metal oxide doped GeO$_2$ for gate stack formation on germanium. *J. Appl. Phys.* **2014**, *116*, 174103. [CrossRef]

6.   Pourtois, G.; Houssa, M.; Delabie, A.; Conard, T.; Caymax, M.; Meuris, M.; Heyns, M.M. Ge 3d core-level shifts at (100)Ge/Ge(Hf)O$_2$ interfaces: A first-principles investigation. *Appl. Phys. Lett.* **2008**, *92*, 032105. [CrossRef]

7.   Sun, J.; Lu, J. Interface engineering and gate dielectric engineering for high performance Ge MOSFETs. *Adv. Cond. Matter Phys.* **2015**, *2015*, 639218. [CrossRef]

8.   Takagi, S.; Kim, S.H.; Yokoyama, M.; Nishi, K.; Zhang, R.; Takenaka, M. Material challenges and opportunities in Ge/III-V channel MOSFETs. *ECS Trans.* **2014**, *64*, 99–110. [CrossRef]

9.   Hong, M.; Kwo, J.; Lin, T.D.; Huang, M.L. InGaAs metal oxide semiconductor devices with Ga$_2$O$_3$(Gd$_2$O$_3$) high-k dielectrics for science and technology beyond Si CMOS. *MRS Bull.* **2009**, *34*, 514–521. [CrossRef]

10. Lin, T.D.; Chang, Y.H.; Lin, C.A.; Huang, M.L.; Lee, W.C.; Kwo, J.; Hong, M. Realization of high-quality HfO$_2$ on In$_{0.53}$Ga$_{0.47}$As by *in-situ* atomic-layerdeposition. *Appl. Phys. Lett.* **2012**, *100*, 172110.

11. Hong, M.; Wan, H.W.; Lin, K.Y.; Chang, Y.C.; Chen, M.H.; Lin, Y.H.; Lin, T.D.; Pi, T.W.; Kwo, J. Perfecting the Al$_2$O$_3$/In$_{0.53}$Ga$_{0.47}$As interfacial electronic structure in pushing metal-oxide-semiconductor field-effect-transistor device limits using *in-situ* atomic-layer-deposition. *Appl. Phys. Lett.* **2017**, *111*, 123502. [CrossRef]

12. Lin, T.D.; Chiu, H.C.; Chang, P.; Tung, L.T.; Chen, C.P.; Hong, M.; Kwo, J.; Tsai, W.; Wang, Y.C. High-performance self-aligned inversion-channel In$_{0.53}$Ga$_{0.47}$As metal-oxide-semiconductor field-effect-transistor with Al$_2$O$_3$/Ga$_2$O$_3$(Gd$_2$O$_3$) as gate dielectrics. *Appl. Phys. Lett.* **2008**, *93*, 033516. [CrossRef]

13. Ren, F.; Kuo, J.M.; Hong, M.; Hobson, W.S.; Lothian, J.R.; Lin, J.; Tsai, H.S.; Mannaerts, J.P.; Kwo, J.; Chu, S.N.G.; et al. Ga$_2$O$_3$(Gd$_2$O$_3$)/InGaAs enhancement-mode n-channel MOSFETs. *IEEE Electron Device Lett.* **1998**, *19*, 309–311. [CrossRef]

14. Sun, S.; Sun, Y.; Liu, Z.; Lee, D.-I.; Peterson, S.; Pianetta, P. Surface termination and roughness of Ge(100) cleaned by HF and HCl solutions. *Appl. Phys. Lett.* **2006**, *88*, 021903. [CrossRef]

15. Kim, H.; McIntyre, P.C.; Chui, C.O.; Saraswat, K.C.; Cho, M.-H. Interfacial characteristics of HfO$_2$ grown on nitrided Ge(100) substrates by atomic-layer deposition. *Appl. Phys. Lett.* **2004**, *85*, 2902. [CrossRef]

16. Lee, W.C.; Chin, B.H.; Chu, L.K.; Lin, T.D.; Lee, Y.J.; Tung, L.T.; Lee, C.H.; Hong, M.; Kwo, J. Molecular beam epitaxy-grown Al$_2$O$_3$/HfO$_2$ high-k dielectrics for germanium. *J. Cryst. Growth* **2009**, *311*, 2187–2190. [CrossRef]

17. Chu, L.K.; Lin, T.D.; Huang, M.L.; Chu, R.L.; Chang, C.C.; Kwo, J.; Hong, M. Ga$_2$O$_3$(Gd$_2$O$_3$) on Ge without interfacial layers: Energy-band parameters and metal oxide semiconductor devices. *Appl. Phys. Lett.* **2009**, *94*, 202108. [CrossRef]

18. Delabie, A.; Bellenger, F.; Houssa, M.; Conard, T.; Elshocht, S.V.; Caymax, M.; Heyns, M.; Meuris, M. Effective electrical passivation of Ge(100) for high-k gate dielectric layers using germanium oxide. *Appl. Phys. Lett.* **2007**, *91*, 082904. [CrossRef]

19. Xie, Q.; Deng, S.; Schaekers, M.; Lin, D.; Caymax, M.; Delabie, A.; Jiang, Y.; Qu, X.; Dedutytsche, D.; Detavernier, C. High-performance Ge MOS capacitors by O$_2$ plasma passivation and O$_2$ ambient annealing. *IEEE Electron Device Lett.* **2011**, *32*, 1656. [CrossRef]

20. Lee, C.H.; Nishimura, T.; Nagashio, K.; Kita, K.; Toriumi, A. High-electron-mobility Ge/GeO$_2$ n-MOSFETs with two-step oxidation. *IEEE Trans. Electron Devices* **2011**, *58*, 1295.

21. Pi, T.W.; Cheng, C.P.; Wertheim, G.K. The reaction of Si(001)-2 × 1 with magnasium. *J. Appl. Phys.* **2011**, *109*, 043701.

22. Pi, T.W.; Hong, I.H.; Cheng, C.P.; Wertheim, G.K. Surface photoemission from Si(100) and inelastic electron mean-free-path in silicon. *J. Electron Spectrosc. Relat. Phenom.* **2000**, *107*, 163–176. [CrossRef]

23. Pi, T.W.; Wen, J.F.; Ouyang, C.P.; Wu, R.T. Surface core-level shifts of Ge(100)-2 × 1. *Phys. Rev. B* **2001**, *63*, 153310. [CrossRef]

24. Pi, T.W.; Wen, J.F.; Ouyang, C.P.; Wu, R.T.; Wertheim, G.K. Oxidation of Si(001)-2 × 1. *Surf. Sci.* **2001**, *478*, L333–L338. [CrossRef]

25. Wada, A.; Zhang, R.; Takagi, S.; Samukawa, S. Formation of thin germanium dioxide film with a high-quality interface using a direct neutral beam oxidation process. *Jpn. J. App. Phys.* **2012**, *51*, 125603. [CrossRef]

26. Sahari, S.K.; Murakami, H.; Fujioka, T.; Bando, T.; Ohta, A.; Makihara, K.; Higashi, S.; Miyazaki, S. Native oxidation growth on Ge(111) and (100) surfaces. *Jpn. J. App. Phys.* **2011**, *50*, 04DA12. [CrossRef]

27. Prabhakaran, K.; Maeda, F.; Watanabe, Y.; Ogino, T. Distinctly different thermal decomposition pathways of ultrathin oxide layer on Ge and Si surfaces. *Appl. Phys. Lett.* **2000**, *76*, 2244. [CrossRef]

28. Milojevic, M.; Contreras-Guerrero, R.; Lopez-Lopez, M.; Kim, J.; Wallace, R.M. Characterization of the "clean-up" of the oxidized Ge(100) surface by atomic layer deposition. *Appl. Phys. Lett.* **2009**, *95*, 212902. [CrossRef]

29. Molle, A.; Spiga, S.; Fanciulli, M. Stability and interface quality of GeO₂ films grown on Ge by atomic oxygen assisted deposition. *J. Chem. Phys.* **2008**, *129*, 011104. [CrossRef]

30. Molle, A.; Bhuiyan, M.N.K.; Tallarida, G.; Fanciulli, M. *In situ* chemical and structural Investigations of the oxidation of Ge(001) substrates by atomic oxygen. *Appl. Phys. Lett.* **2006**, *89*, 083504. [CrossRef]

31. Yoshigoe, A.; Teraoka, Y.; Okada, R.; Yamada, Y.; Sasaki, M. *In situ* synchrotron radiation photoelectron spectroscopy study of the oxidation of the Ge(001)-2 × 1 surface by supersonic molecular oxygen beams. *J. Chem. Phys.* **2014**, *141*, 174708. [CrossRef]

32. Cho, J.H.; Kim, K.S.; Morikawa, Y. Structure and binding energies of unsaturated hydrocarbons on Si(001) and Ge(001). *J. Chem. Phys.* **2006**, *124*, 024716. [CrossRef]

33. Fan, X.L.; Lau, W.M.; Liu, Z.F. Nondissociative Adsorption of O₂ on Ge(100). *J. Phys. Chem. C* **2009**, *113*, 8786–8793. [CrossRef]

34. Fleischmann, C.; Schouteden, K.; Merckling, C.; Sioncke, S.; Meuris, M.; Haesendonck, C.V.; Temst, K.; Vantomme, A. Adsorption of O₂ on Ge(100): Atomic geometry and site-specific electronic structure. *J. Phys. Chem. C* **2012**, *116*, 9925–9929. [CrossRef]

35. Pi, T.W.; Lin, Y.H.; Fanchiang, Y.T.; Chiang, T.H.; Wei, C.H.; Lin, Y.C.; Wertheim, G.K.; Kwo, J.; Hong, M. *In-situ* atomic layer deposition of tri-methylaluminum and water on pristine single-crystal (In)GaAs surfaces: Electronic and electric structures. *Nanotechnology* **2015**, *26*, 164001. [CrossRef]

36. Fukuda, T.; Ogino, T. Initial oxygen reaction on Ge(100) 2 × 1 surfaces. *Phys. Rev. B* **1997**, *56*, 13190–13193. [CrossRef]

37. Shah, G.A.; Radny, M.W.; Smith, P.V. Initial stages of oxygen chemisorption on the Ge(001) surface. *J. Phys. Chem. C* **2014**, *118*, 15795–15803. [CrossRef]

38. Soon, J.M.; Lim, C.W.; Loh, K.P.; Ma, N.L.; Wu, P. Initial-stage oxidation mechanism of Ge(100) 2 × 1 dimers. *Phys. Rev. B* **2005**, *72*, 115343. [CrossRef]

39. Franco, J.; Kaczer, B.; Roussel, P.J.; Mitard, J.; Sioncke, S.; Witters, L.; Mertens, H.; Grasser, T.; Groeseneken, G. Understanding the suppresed charge trapping in relaxed- and strained-Ge/SiO₂/HfO₂ pMOSFETs and implications for the screening of alternative high-mobility substrate/dielectric CMOS gate stacks. In Proceedings of the 2013 IEEE International Electronic Devices Meeting, Washington, DC, USA, 9–11 December 2013; pp. 358–362.

40. Wan, H.W.; Hong, Y.J.; Cheng, Y.T.; Kwo, J.; Hong, M. BTI Characterization of MBE Si-capped Ge gate stack and defect reduction via forming gas annealing. In Proceedings of the 2019 IEEE International Reliability Symposium, Monterey, CA, USA, 31 March–4 April 2019.

41. Lee, K.Y.; Lee, W.C.; Lee, Y.J.; Huang, M.L.; Chang, C.H.; Wu, T.B.; Hong, M.; Kwo, J. Molecular beam epitaxy grown template for subsequent atomic layer deposition of high κ dielectrics. *Appl. Phys. Lett.* **2006**, *89*, 222906. [CrossRef]

42. Cheng, Y.T.; Lin, Y.H.; Chen, W.S.; Lin, K.Y.; Wan, H.W.; Cheng, C.P.; Cheng, H.H.; Kwo, J.; Hong, M.; Pi, T.W. Surface electronic structure of epi Germanium (001)-2 × 1. *Appl. Phys. Express* **2017**, *10*, 075701. [CrossRef]

43. Cao, R.; Yang, X.; Terry, J.; Pianetta, P. Core-level shifts of the Ge(100)-(2 × 1) surface and their origins. *Phys. Rev. B* **1992**, *45*, 13749–13752. [CrossRef]

44. Eriksson, P.E.J.; Uhrberg, R.I.G. Surface core-level shifts on clean Si(001) and Ge(001) studied with photoelectron spectroscopy and density functional theory calculations. *Phys. Rev. B* **2010**, *81*, 125443. [CrossRef]

45. Goldoni, A.; Modesti, S.; Dhanak, V.R.; Sancrotti, M.; Santoni, A. Evidence for three surface components in the 3d core-level photoemission spectra of Ge(100)-(2 × 1) surface. *Phys. Rev. B* **1996**, *54*, 11340–11345. [CrossRef]

46. Landemark, E.; Karlsson, C.J.; Johansson, L.S.O.; Uhrberg, R.I.G. Electronic structure of clean and hydrogen-chemisorbed Ge(001) surfaces studied by photoelectron spectroscopy. *Phys. Rev. B* **1994**, *49*, 16523–16533. [CrossRef]

47. Le Lay, G.; Kanski, J.; Nilsson, P.O.; Karlsson, U.O.; Hricovini, K. Surface core-level shifts on Ge(100): c(4 × 2) to 2 × 1 and 1 × 1 phase transitions. *Phys. Rev. B* **1992**, *45*, 6692–6699. [CrossRef]
48. Margaritondo, G.; Rowe, J.E.; Christman, S.B. Photoionization cross section of d-core levels in solids: A synchrotron radiation study of the spin-orbit branching ratio. *Phys. Rev. B* **1979**, *19*, 2850–2855. [CrossRef]
49. Aarts, J.; Hoeven, A.-J.; Larsen, P.K. Core-level study of the phase transition on the Ge(111)-c(2 × 8) surface. *Phys. Rev. B* **1988**, *38*, 3925–3930. [CrossRef]
50. Pehlke, E.; Scheffler, M. Evidence for site-sensitive screening of core holes at the Si and Ge (001) surface. *Phys. Rev. Lett.* **1993**, *71*, 2338–2341. [CrossRef] [PubMed]
51. Binder, J.F.; Broqvist, P.; Komsa, H.; Pasquarello, A. Germanium core-level shifts at Ge/GeO$_2$ interfaces through hybrid functionals. *Phys. Rev. B* **2012**, *85*, 245305. [CrossRef]
52. Schmeisser, D.; Schnell, R.D.; Bogen, A.; Himpsel, F.J.; Rieger, D.; Landgren, G.; Morar, J.F. Surface oxidation states of germanium. *Surf. Sci.* **1986**, *172*, 455–465. [CrossRef]
53. Hansen, D.A.; Hudson, J.B. Oxygen scattering and initial chemisorption probability on Ge(100). *Surf. Sci.* **1991**, *254*, 222–234. [CrossRef]
54. Cheng, C.P.; Chen, W.S.; Lin, K.Y.; Wei, G.J.; Cheng, Y.T.; Lin, Y.H.; Wan, H.W.; Pi, T.W.; Tung, R.T.; Kwo, J.; et al. Atomic nature of the Schottky barrier height formation of the Ag/GaAs(001)-2 × 4 interface: An in-situ synchrotron radiation photoemission study. *Appl. Surf. Sci.* **2017**, *393*, 294–298. [CrossRef]
55. Kronik, L.; Shapira, Y. Surface photovoltage phenomena: Theory, experiment, and applications. *Surf. Sci. Rep.* **1999**, *37*, 1–206. [CrossRef]
56. Radny, M.W.; Shah, G.A.; Schofield, S.R.; Smith, P.V.; Curson, N.J. Valence surface electronic states on Ge(001). *Phys. Rev. Lett.* **2008**, *100*, 246807. [CrossRef]

nanomaterials

MDPI

Article

In Situ X-ray Photoelectron Spectroscopic and Electrochemical Studies of the Bromide Anions Dissolved in 1-Ethyl-3-Methyl Imidazolium Tetrafluoroborate

Jaanus Kruusma [1], Arvo Tõnisoo [2], Rainer Pärna [2], Ergo Nõmmiste [2] and Enn Lust [1,*

[1]   Institute of Chemistry, University of Tartu, Ravila 14A, 50411 Tartu, Estonia; jaanus.kruusma@ut.ee
[2]   Institute of Physics, University of Tartu, W. Ostwaldi 1, 50411 Tartu, Estonia; arvo.tonisoo@ut.ee (A.T.); rainer.parna@ut.ee (R.P.); ergo.nommiste@ut.ee (E.N.)
*   Correspondence: enn.lust@ut.ee; Tel.: +372-737-5165

Received: 22 January 2019; Accepted: 18 February 2019; Published: 22 February 2019

Abstract: Influence of electrode potential on the electrochemical behavior of a 1-ethyl-3-methylimidazolium tetrafluoroborate (EMImBF$_4$) solution containing 5 wt % 1-ethyl-3-methylimidazolium bromide (EMImBr) has been investigated using electrochemical and synchrotron-initiated high-resolution in situ X-ray photoelectron spectroscopy (XPS) methods. Observation of the Br $3d_{5/2}$ in situ XPS signal, collected in a 5 wt % EMImBr solution at an EMImBF$_4$–vacuum interface, enabled the detection of the start of the electrooxidation process of the Br$^-$ anion to Br$_3^-$ anion and thereafter to the Br$_2$ at the micro-mesoporous carbon electrode, polarized continuously at the high fixed positive potentials. A new photoelectron peak, corresponding to B–O bond formation in the B 1s in situ XPS spectra at $E \leq -1.17$ V, parallel to the start of the electroreduction of the residual water at the micro-mesoporous carbon electrode, was observed and is discussed. The electroreduction of the residual water caused a reduction in the absolute value of binding energy vs. potential plot slope twice to ca. $dBE\, dE^{-1} = -0.5$ eV V^{-1} at $E \leq -1.17$ V for C 1s, N 1s, B 1s, F 1s, and Br $3d_{5/2}$ photoelectrons.

Keywords: room temperature ionic liquids; in situ X-ray photoelectron spectroscopy; binding energies; cyclic voltammetry; electrochemical impedance spectroscopy; micro-mesoporous carbon electrode; supercapacitor materials

1. Introduction

Electricity is one of the most convenient modes of energy that can be very easily converted into other forms of energy. Besides the flexibility, the electrical energy does not pollute the surrounding environment and the electrical devices are small and quiet. Therefore, electricity has found applications in many fields of modern technology including electrochemical power sources, electrosynthesis, and galvanic processes. In mobile applications and isolated places, electrical devices should have high specific energy and power density. In terms of environmental protection and sustainability, i.e., the recycle economy point of view, electrical energy generating devices should be reusable, i.e., rechargeable.

Two types of reusable powerful electrical energy storage systems are known and applied: rechargeable electrochemical faradic cells and supercapacitors [1–11]. Supercapacitors are characterized by very high specific power density and capacitance values (up to 175 F g^{-1} for aqueous and up to 100 F g^{-1} for nonaqueous electrolyte-based, commercial electrochemical double layer capacitor (EDLC) cells [1–3], and from 120 to 150 F g^{-1} for novel micro-mesoporous carbon electrode-based systems in nonaqueous electrolytes [11–18]). The number of recharging cycles that can be applied exceeds 500,000, more than 100 times higher than the corresponding number of rechargeable electrochemical cells [1,19,20]. However, the energy density of supercapacitors is much lower than that for faradic electrochemical cells [1,6]. Therefore, it seems to be very attractive to combine the superior properties of both types of electrical energy storage systems into a common device, a so-called hybrid capacitor, where the electrical double layer charging is combined with the fast reversible redox process(es) taking place at the faradic-type electrode [1,3,4]. Different redox couples (e.g. hydroxyquinone/quinone, Ru/RuO_2, $MnO(OH)/MnO_2$, I^-/I_2, metal chalcogenides, etc.) have been studied for the construction of hybrid capacitors [1,3,4]. Various electrolytes, based on aqueous or non-aqueous solutions, liquid or solid, have been tested as well [1,3,4,19–29]. It should be noted that the applicable cell potential (ΔE) is moderate and limited (ca. 1.2 ... 1.6 V) for the aqueous electrolyte-based hybrid capacitors due to electrochemical decomposition of water at $\Delta E \geq 1.2$ V [3,20,21]. Therefore, non-aqueous electrolytes, having a wider applicable ΔE range, are more desirable for the construction of hybrid supercapacitors due to the higher energy density stored [1–3].

The properties of iodide anions containing non-aqueous electrolyte systems have been studied in various electrodes [30–36]. Remarkably, high series capacitance values (more than 120 mF cm^{-2}) have been measured for pyrolytic graphite — 5 wt % 1-ethyl-3-methylimidazolium iodide (EMImI) dissolved in a 1-ethyl-3-methylimidazolium tetrafluoroborate (EMImBF$_4$) interface ($E > 0.5$ V vs. Ag/AgCl) [33]. Very high specific capacitance values (up to 245 F g^{-1} at the $\Delta E = 1.0$ V) have been measured for a D-glucose-derived, micro-mesoporous carbon-based EDLC [35]. However, for a 5 wt % EMImI solution in EMImBF$_4$, the nearly reversible I$^-$ anion adsorption takes place only at $\Delta E \leq 2.4$ V. Within the cell potential range from 2.6 to 3.0 V, the complicated mixed kinetic faradic processes dominate, decreasing the EDLC reversibility and the energetic efficiency of the device [35].

Yamazaki et al. [37] investigated the capacitive properties of 1.0 M 1-ethyl-3-methylimidazolium bromide (EMImBr) dissolved in EMImBF$_4$ at the activated carbon fiber cloth electrode. Charging this system up to $\Delta E = 2.0$ V (at the gravimetric current density (i_g) $i_g = 100$ mA g^{-1}), a 59.0 F g^{-1} specific capacitance was obtained. It is obvious that this specific capacitance value is much lower than that for the I$^-$ anion-based system, as the specific adsorption of Br$^-$ anion is much weaker than that of the I$^-$ anion [35] However, the bromide anion-containing system possessed excellent cycleability and a coulombic efficiency up to $\Delta E = 2.0$ V [37].

Adsorption of the Br$^-$ anion dissolved in EMImBF$_4$ at the Bi(111) electrode has demonstrated better reversibility compared to I$^-$ anion adsorption [38]. These data indicate that the hybrid EDLCs filled with non-aqueous electrolytes containing the Br$^-$ anion could have higher energy efficiencies than devices based on the I$^-$ anion.

Quite recently, Gastrol et al. [39] published a work where the capacitance of the bromide anion containing a hybrid EDLC was extended up to 314 F g^{-1} at $\Delta E = 1.1$ V by the partial oxidation of Br$^-$ anion (in 2 M KOH aq. solution) to the BrO$_3^-$ anion. However, the ΔE for the studied system was limited by the start of the electrochemical decomposition of water at the positively charged activated carbon electrode [39].

Therefore, due to the limited amount of information characterizing the electrochemical properties of the bromide anion containing room temperature ionic liquid electrolytes at micro-mesoporous carbon electrodes, the micro-mesoporous molybdenum carbide-derived carbon (C(Mo$_2$C) electrode in a 5 wt % EMImBr solution in EMImBF$_4$ was studied using cyclic voltammetry (CV), electrochemical impedance spectroscopy (EIS), and synchrotron radiation-initiated high-resolution in situ X-ray photoelectron spectroscopy (XPS) methods.

2. Materials and Methods

The in situ XPS spectra were recorded at the polarized Mo$_2$C derived carbon (C(Mo$_2$C)) electrodes under high vacuum at the synchrotron-initiated adjustable energy X-ray beamline I411, Max II Laboratory, Lund University (Lund, Sweden).

C(Mo$_2$C) electrodes (the working electrode, WE), covered with a very thin layer of a 5 wt % EMImBr ($\geq$99%, Iolitec Ionic Liquids Technologies, Heilbronn, Germany) solution in EMImBF$_4$

($\geq$99.0% (HPLC), Fluka, Honeywell, Bucharest, Romania), containing less than 200 ppm water, were studied. The lower part of the WE was soaked and kept during the XPS measurement in room temperature ionic liquids (RTILs) containing a reservoir. The 5 wt % EMImBr solution in EMImBF$_4$ was prepared and stored in an Ar-filled glove box containing less than 0.1 ppm water and oxygen. The C(Mo$_2$C) electrode was positioned almost vertically, polarized during the in situ XPS experiments using a three-electrode electrochemical cell. Platinum gauze (with an apparent area of ca. 2 cm^2 and a grid size of 100 mesh, 99.9%, Merck KGaA, Darmstadt, Germany) was used as the counter electrode (CE), and Ag wire covered with AgCl (Ag/AgCl in pure EMImBF$_4$ RTIL) was used as the reference electrode (RE) in RTIL.

The design of the synchrotron radiation beamline, the working, counter, and reference electrodes, and the used electrochemical cell have been described in detail [40]. The reproducibility data for the synchrotron radiation-initiated in situ XPS measurements for C 1s, N 1s, F 1s, and B 1s signals have been published as well [40]. The stability of the binding energy (*BE*) peak value for the C 1s signal was $\pm$0.12 eV. The number of independent experiments was 17, and the relative standard deviation was only $\pm$0.041% over 9.5 h of test time [41]. The measured amounts of detected photoelectrons, forming XPS peaks, had the following stabilities: for aliphatic carbon (C$_5$, Figure 1) C 1s, RSD = $\pm$8.8 %, for B 1s, RSD = $\pm$ 3.4 %, for N 1s, RSD = $\pm$ 6.2 %, and for F 1s, RSD = $\pm$7.6 % [40]. For additional supporting electrochemical measurements (inside the very dry Ar-filled glove-box), a carbon fiber (*d* = 11 µm, Bioanalytical Systems, Inc., West Lafayette, IN, USA) microelectrode was used.

For a better understanding and a more detailed discussion of the collected in situ XPS data, the elements, forming the EMIm$^+$ cation, were numbered, as shown in Figure 1. XPS data analysis and *BE* spectrum fitting procedures were performed using IgorPro (ver. 6.2.2.2, WaveMetrics, Inc., Lake Oswego, OR, USA) and CasaXPS (ver. 2.315, Casa Software Ltd., Teignmouth, UK) software, respectively [40]. The C 1s XPS spectra were fitted using the four photoelectron (PE) peak model, where one C 1s PE peak was related to the aliphatic carbon (C$_5$, Figure 1), and the other three C 1s PE peaks were related to the hetero-aromatic ring (included or bounded) carbons (C$_1$, C$_2$, C$_3$, C$_4$, and C$_6$, as noted in Figure 1) in the same manner as it was described earlier by Foelske-Schmitz et al. [42] and Licence et al. [43,44] using a combined Gaussian–Lorentz function with ratio 70:30, respectively. The full width at half maximum (FWHM) of the C 1s PE peak, related to a so-called hetero-aromatic carbon, was allowed to change between 0.9 and 1.1 eV, as described by Licence et al. [43]. The FWHM for the C 1s PE peak related to aliphatic carbon (C$_5$) was unfixed. The other three PE peak positions were fixed relative to each other. The *BE* of C$_2$ was equal to the *BE* of C$_3$, and the *BE* of C$_4$ was equal to the *BE* of C$_6$. *BE* separation between C$_2$ and C$_1$ was fixed to 0.70 eV, and *BE* separation between C$_2$ and C$_4$ was fixed to 0.50 eV. The *BE* of C$_5$, representing the carbon atoms in the alkyl chain (having sp^3 electronic configuration), was fixed to 285.3 eV based on the non-polarized C(Mo$_2$C) WE data. The PE peak areas for C$_1$, C$_2$+C$_3$, and C$_4$+C$_6$ carbons were fixed at a ratio of 0.5:1:1 [42].

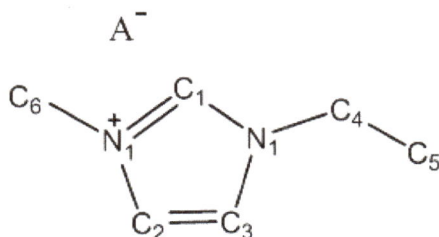

Figure 1. Notation of the carbon and nitrogen atoms of 1-ethyl-3-methylimidazolium (EMIm$^+$) cation. A$^-$ marks the Br$^-$ or BF$_4^-$ anion, respectively.

The N 1s, B 1s, F 1s, and Br 3d X-ray photoelectron spectra were fitted using the same ratio of the Gaussian–Lorentz function as described for C 1s XPS, leaving the FWHM and peak positions free. Later, the obtained individual peak *BE* was corrected according to the *BE* of the C_5 carbon measured at fixed potential.

The ratios of the XPS PE peaks have been calculated using the following equation:

$$X_{x}, \% = 100 \times \frac{A_x}{\sum\limits_{i=1}^{n} A_i}$$

where X_x is the ratio of the PE peak "*x*," and A_x is the amount of the counted PE (i.e., the PE peak area) of the XPS signal "*x*," corrected with the synchrotron ring current value and the number of scans.

In addition to XPS studies, cyclic voltammetry (CV) and electrochemical impedance spectroscopy (EIS) measurements were performed in the oxygen and humidity-free Ar-filled glovebox. A Gamry Instruments "Reference 3000" potentiostat, controlled by the Gamry Instruments "Framework" (ver. 6.32) software, was used to polarize the electrodes. The potential sweep rate of 1.0 mV s^{-1} and the potential step of 5 mV were used for obtaining the CVs. The EIS measurements were performed in potentiostatic mode (vs. the RE potential), and the ac modulation amplitude was 5.0 mV. A single sinusoidal potential wave was used for modulation of the C(Mo_2C) electrode potential. The XPS and electrochemical measurements were conducted at 22 °C.

3. Results and Discussion

3.1. Characteristic Changes in C 1s, N 1s, B 1s, F 1s, and Br 3d In Situ X-ray Spectra Obtained at the Negatively Polarized C(Mo$_2$C) Electrode

The X-ray photoelectron spectra for C 1s, N 1s, B 1s, F 1s, and Br 3d PEs at fixed negative potentials were recorded within the potential range from −2.07 to −0.27 V and are presented in the Figure 2a–e, respectively. Exact positions of the PE peaks are presented in Table S1. The PE spectra indicate significant changes for the aliphatic carbon (C_5) C 1s (Figures 2a and 3) and nitrogen (N1) N 1s (Figures 2b and 3) XPS signals at E = −1.27 V. For N 1s XPS, a very small new PE peak formed at E = −0.67 V (located at *BE* = 400.3 eV and marked hereafter as N2, not shown for space constraints), causing a very small deviation in the relative amount of N1 1s signal (Figure 3). A shift of the C(Mo_2C) electrode potential toward more negative values caused a noticeable increase in the relative size of the N2 1s PE peak. At E = −1.17 V, a small shoulder formed in the initial B 1s PE peak (at *BE* = 194.8 eV, defined as B1 hereafter and in Figures 2c and 3). Parallel to this, gas bubbles started to form occasionally at the C(Mo_2C) working electrode surface. This shoulder corresponded to a new B 1s PE peak (defined as B2 hereafter and in Figure 2c) with *BE* = 193.3 eV. The initial B 1s XPS peak (B1) was located at *BE* = 194.8 eV (E = −1.17 V), based on the NIST XPS database [45]. The new B 1s PE peak (B2) could correspond to the formation of some sort of boron–oxygen compound.

No changes in the shapes of the F 1s (Figures 2d and 3) and Br 3d (Figures 2e and 3) XPS signals were noted, indicating high electrochemical stability of these elements within the negative electrode potential range. The decrease in intensities of the B 1s, F 1s, and Br 3d XPS signals, notable in Figure 2c–e, could be explained by the decrease in the concentration of Br^- and BF_4^- anions at the ionic liquid–carbon surface due to electrostatic repulsion of anions from the negatively charged C(Mo_2C) electrode.

Analysis of the relationship between the C_5 1s, N1 1s, B1 1s, F 1s, and Br $3d_{5/2}$ PE BEs upon the potential applied to micro-mesoporous C(Mo_2C) electrode indicated that $dBE\ dE^{-1} = -1$ eV V^{-1} for C_5 1s, N1 1s, B1 1s, F 1s, and Br $3d_{5/2}$ within the potential range from −1.17 to 1.23 V (the $dBE\ dE^{-1}$ slope values are presented in Table S2, given in the supplementary information). However, for the potential range from −2.07 to −1.17 V, $dBE\ dE^{-1} = -0.5$ eV V^{-1} for C_5 1s, N1 1s, B1 1s, F 1s, and Br $3d_{5/2}$ PEs. Therefore, the twofold decrease in the $dBE\ dE^{-1}$ slope, the formation of the gas bubbles at the C(Mo_2C) electrode and the new shoulder into the initial B 1s PE peak, and the formation of a B–O bond at E = −1.17 V were initiated by reduction processes at the electrode surface.

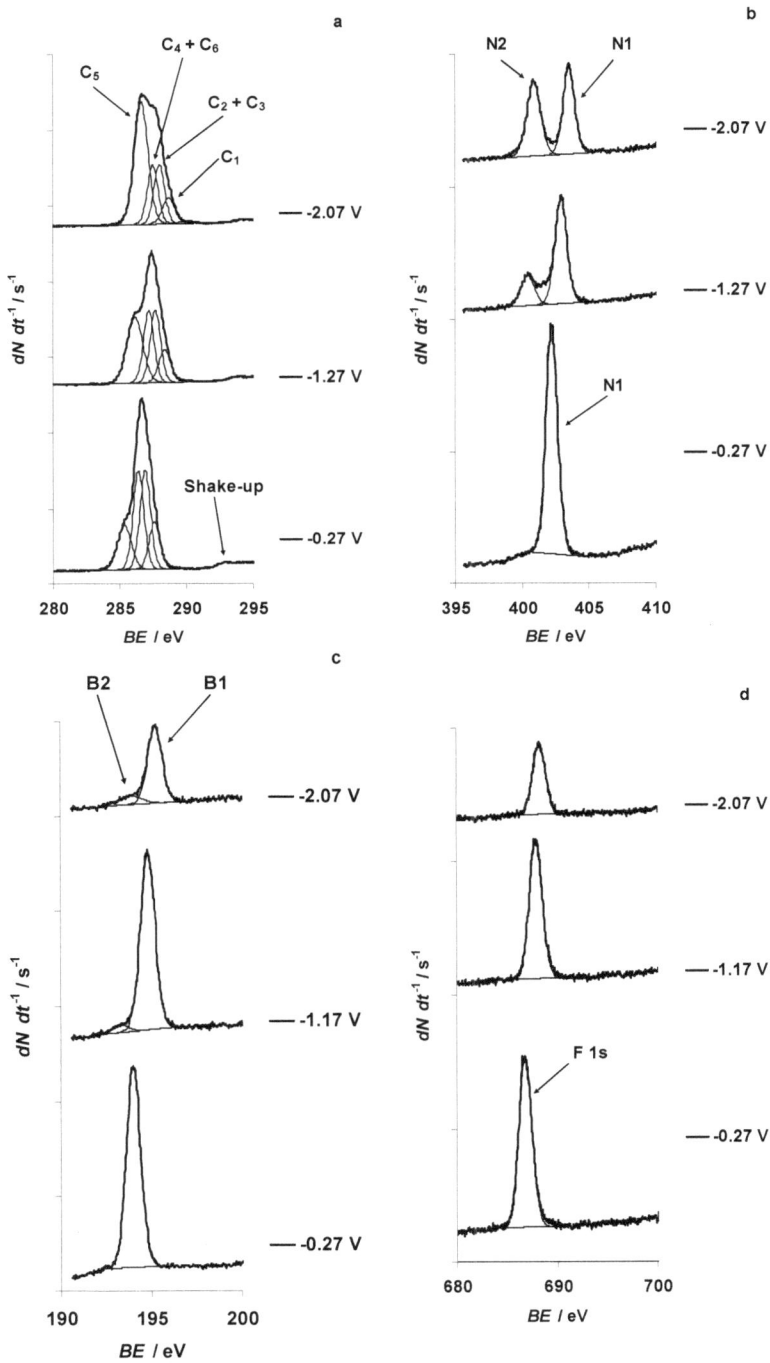

Figure 2. *Cont.*

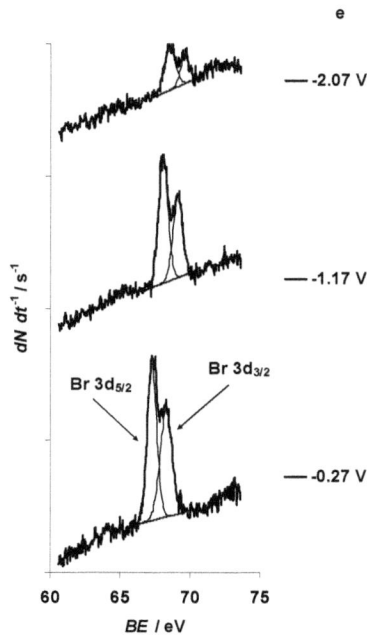

Figure 2. C 1s, N 1s, B 1s, F 1s, and Br 3d X-ray photoelectron (PE) spectra for the 5 wt % EMImBr solution in EMImBF$_4$ measured at the fixed negative potentials of the C(Mo$_2$C) electrode. The binding energy (*BE*) scales for the XPS experiments have been referenced to the *BE* of the C 1s photoemission line related to aliphatic carbon (*BE* = 285.3 eV) measured for the non-polarized and grounded electrode. (**a**) C 1s PE spectra, fitted applying four C 1s photoelectron peak model (regular lines) at various potentials noted in the figure; excitation energy was 400 eV, and the PE signal intensity (d*N* d*t*$^{-1}$) scale between tick marks was 100 counts s^{-1}. (**b**) N 1s PE spectra; excitation energy was 500 eV, and the PE signal intensity scale between tick marks was 35 counts s^{-1}. (**c**) B 1s PE spectra; excitation energy was 250 eV, and the PE signal intensity scale between tick marks was 20 counts s^{-1}. (**d**) F 1s PE spectra; excitation energy was 800 eV, and the PE signal intensity scale between tick marks was 8 counts s^{-1}. (**e**) Br 3d PE spectra; excitation energy was 120 eV, and the PE signal intensity scale between tick marks was 20 counts s^{-1}.

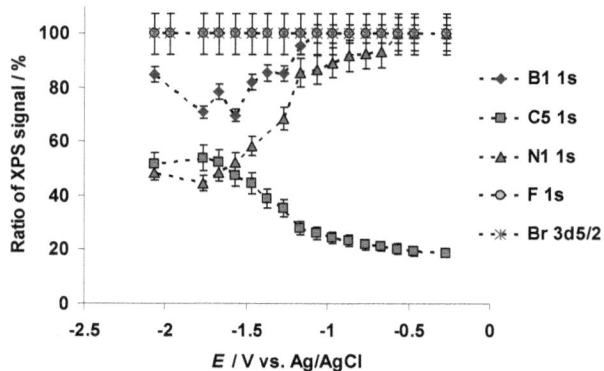

Figure 3. Dependence of EMIm$^+$ "aliphatic" carbon C 1s (marked in the figure as C5), imidazolium nitrogen atoms N 1s (N1), BF$_4$$^-$ anion boron B 1s (B1), BF$_4$$^-$ anion fluorine F 1s, and Br$^-$ anion bromine Br 3d$_{5/2}$ photoelectron peaks ratios for the 5 wt % EMImBr solution in EMImBF$_4$ at various C(Mo$_2$C) electrode negative potentials.

3.2. The Electrochemical Measurements Data at Negatively Polarized C(Mo₂C) Electrode

The cyclic voltammetry (CV) curves for the negatively polarized micro-mesoporous C(Mo₂C) electrode soaked in the 5 wt % EMImBr solution in EMImBF₄ at various fixed potential sweep regions are shown in Figure 4. Intensive electroreduction of the imidazolium cation [13,40,46,47] started at $E = -1.90$ V, parallel to the remarkable increase in the pressure in the XPS measurement chamber (Figure 5a). (The behavior of the XPS chamber pressure at the positive C(Mo₂C) electrode potentials is explained later in the text.) In Figure 5b, the XPS vacuum chamber pressure values, containing also our previously published data [40,46,47], show that intensive electroreduction processes started at comparable negative potentials, depending only weakly on the chemical composition of the anions in the electrolyte solution.

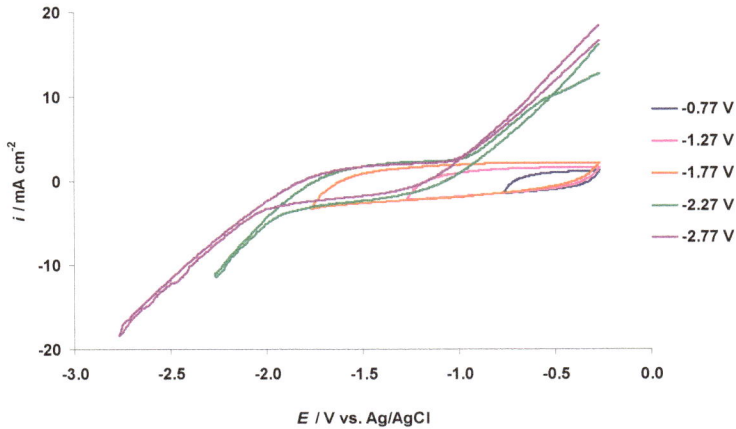

Figure 4. Cyclic voltammetry (CV) curves for negatively polarized C(Mo₂C) electrode in the 5 wt % EMImBr solution in EMImBF₄ measured at variable potential ranges and normal pressure in the Ar-filled glove box conditions (second CV scans are presented starting and ending at $E = -0.27$ V, and the potential scan rate was 1.0 mV s^{-1}).

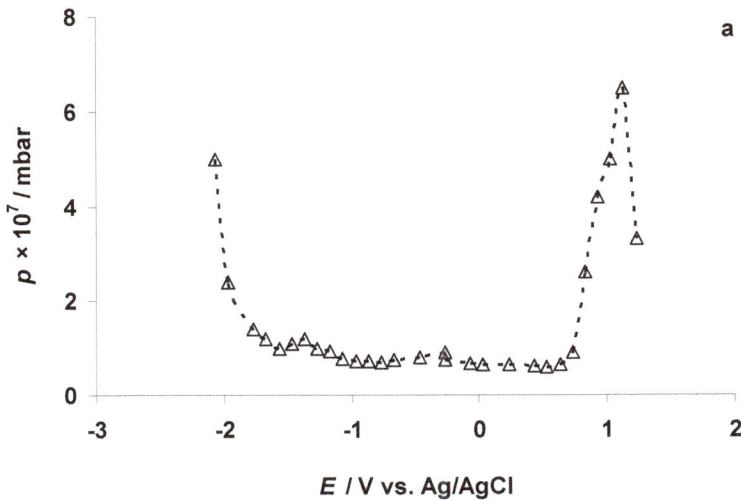

Figure 5. *Cont.*

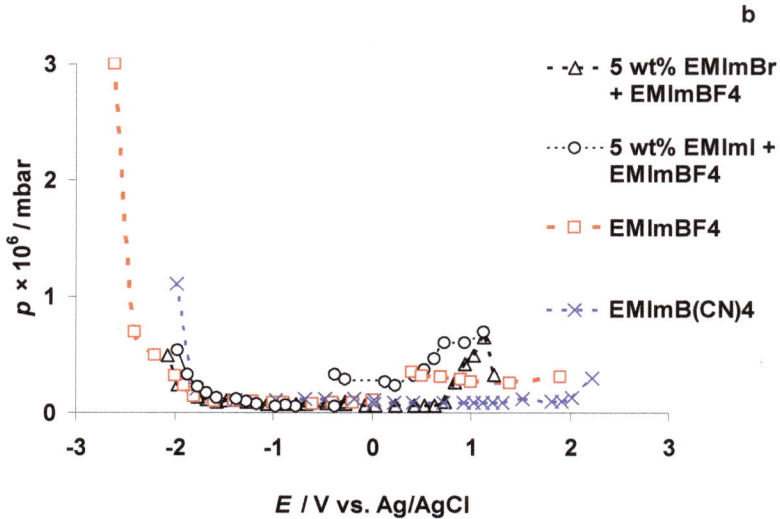

Figure 5. The dependence of gas pressure, *p* (measured inside the vacuum chamber of the X-ray photoelectron spectrometer), on the negative and positive potentials applied to the C(Mo_2C) electrode soaked in (**a**) the 5 wt % EMImBr solution in $EMImBF_4$ and (**b**) the 5 wt % EMImBr solution in $EMImBF_4$ ($\triangle$), $EMImB(CN)_4$ ($\times$), $EMImBF_4$ ($\square$), and the 5 wt % EMImI solution in $EMImBF_4$ ($\bigcirc$).

For more detailed analysis of the electrochemical processes in the 5 wt % EMImBr solution in $EMImBF_4$, potential linear sweep measurements were performed at the carbon fiber microelectrode (d = 11 $\pm$ 2 µm). The carbon fiber microelectrode was selected to suppress the "masking" effect of the electrical double layer charging capacitive current during the potential sweep and to have more effective mass transport of possible reagents to the electrode surface. Data of the second, more negative potential values moving toward sweep shows that the intensive electrochemical reduction process started in the 5 wt % EMImBr solution in the $EMImBF_4$–C interface only at *E* < −1.90 V, where nearly an exponential increase in current density takes place (Figure 6, brown line). The amplified low current density section in Figure 6, brown line, indicates that a slight increase in current density starts at *E* = −1.74 V followed by the most intensive increase at *E* = −1.90 V. This slight increase in the current might indicate the adsorption of the $EMIm^+$ cations before the start of the electroreduction of $EMIm^+$ cations. It is also interesting to note that the reduction current densities at *E* > −1.97 V are much lower for the potential sweep curve collected from −0.27 to −2.27 V than the currents for the potential sweeps performed before, and stopped at *E* ≥ −1.77 V. This phenomenon could be explained by partial passivation of the carbon fiber microelectrode within the potential range −2.27 V < *E* < −1.77 V during the first potential sweep (not shown for clearance) due to the irreversible reduction of $EMIm^+$ cations and the formation of the dielectric EMIm–EMIm dimer film at the carbon electrode [13]. However, for the potential sweeps conducted within the potential ranges from −0.27 to −1.27 V (orange line), from −0.27 to −1.52 V (green line), and from −0.27 to −1.77 V (violet line), there was a gradual electrochemical activation of the 5 wt % EMImBr solution in the $EMImBF_4$–C microelectrode interface (Figure 6).

The d*i* dE^{-1} curve, constructed using the potential linear sweep data collected within the potential range from −0.27 to −1.27 V, shows a d*i* dE^{-1} peak (d*i* dE^{-1} = −16.5 mA cm^{-2} V^{-1}) at *E* = −0.78 V (Figure 7), marking the occurrence of an electroreduction process. The curve passes through a minimum at *E* = −0.89 V. A new, wide negative current maximum has a peak value (−15.0 mA cm^{-2} V^{-1}) at *E* = −1.19 V (Figure 7). This second, very wide negative current maximum is likely caused by the very slow electroreduction of the residual water at the carbon electrode. The electrochemical reduction of the residual water at the glassy carbon electrode was reported by Cheek et al. [48].

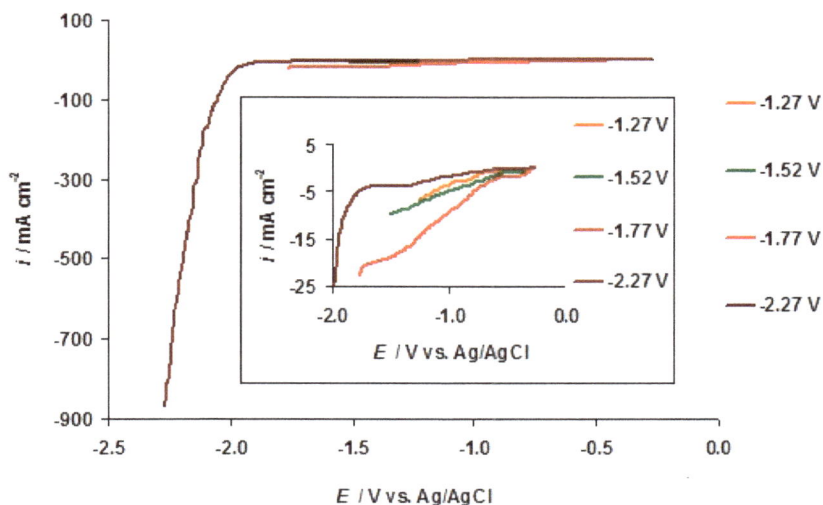

Figure 6. Potential linear sweep (LS) data recorded for negatively polarized C fiber microelectrode soaked in the 5 wt % EMImBr solution in EMImBF$_4$. Second sweeps moving toward more negative potentials are shown, measured at the normal pressure in the Ar-filled glove box conditions. Potential sweeps started at $E = -0.27$ V and ended at the potentials indicated in the figure; the potential scan rate was 1.0 mV s^{-1}. The inset represents the zoomed in part of the LSs shown above.

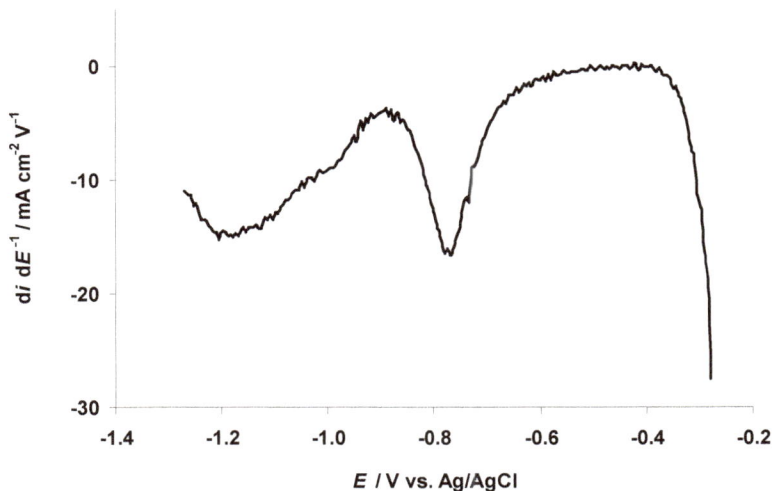

Figure 7. Differentiated potential linear sweep curve (di dE^{-1}) recorded for negatively polarized C fiber microelectrode soaked in the 5 wt % EMImBr solution in EMImBF$_4$. The second sweep was shown measured at the normal pressure in the Ar-filled glove box conditions starting at $E = -0.27$ V and ending at $E = -1.27$ V, with a potential scan rate of 1.0 mV s^{-1}.

The reduction of the dBE dE^{-1} slope to -0.5 eV V^{-1} at $E \leq -1.17$ V for C 1s, N 1s, B 1s, F 1s, and Br 3d$_{5/2}$ PE signals (Table S2), the start of the low intensity formation of gas bubbles (hydrogen formation and evolution) at the C(Mo$_2$C) electrode, and the formation of a new shoulder into the B 1s PE peak (Figures 2c and 3) indicated the formation of a B–O bond at $E = -1.17$ V, initiated by the electrochemical reduction of the residual water (210 ppm, based on the Karl Fisher titration method) in the 5 wt % EMImBr solution in EMImBF$_4$. We suppose that the reduction of dBE dE^{-1} slope to the value ca. -0.5 eV V^{-1} was caused by the chemisorption [49,50] of partially hydrolyzed EMImBF$_4$, which is more hydrophilic due to the B–O bonds formed at the C(Mo$_2$C) electrode surface at $E \leq -1.17$ V (Table S2).

The CV data for the 5 wt % EMImBr solution in the EMImBF$_4$–C microelectrode interface (Figures 6 and 7) are in an agreement with the changes noted for C$_5$ 1s, N 1s, and B 1s XPS data (Figures 2a–c and 3). The formation of a boron–oxygen bond at $E = -1.17$ V and the stability of the B2 1s PE peak at more negative C(Mo$_2$C) potentials confirm the electrochemical reduction of the residual water from the 5 wt % EMImBr solution in EMImBF$_4$.

However, the formation of the B–O bond only at $E \leq -1.17$ V indicated the chemical and electrochemical stability of the BF$_4^-$ anion at less negative potentials in the presence of residual water. Thus, the formation of a new B 1s PE peak at ca. $E = -1.1$ V could be used as an indicator of the presence of the residual water in the electrochemical system.

The electrochemical impedance spectroscopy (EIS) data, i.e., Nyquist plots measured in the potentiostatic regime for the 5 wt % EMImBr solution in the EMImBF$_4$–C(Mo$_2$C) interface, have a stable characteristic shape from $E = -0.27$ to $E = -1.67$ V. Nyquist plots consist of a high-frequency semicircle (caused by the restricted mass transport in the micro-mesoporous C(Mo$_2$C) electrode) followed by a semi-vertical line formed at medium and low frequencies, demonstrating the slow adsorption of the RTIL ions at the energetically inhomogeneous micro-mesoporous C(Mo$_2$C) electrode surface. At $E = -1.77$ V, a small inductive loop forms at the end of the small high frequency semicircle, and the slope of the low frequency line starts to reduce (Figure 8a,b). The Nyquist plots indicate also that, at $E \leq -1.87$ V, the imaginary values of the impedance (Z'') become less negative at low frequencies, and the low-frequency part of the Nyquist plot initially becomes almost parallel with the axis of the real part of the impedance (Z'), indicating the slow charge transfer step-limited process.

At $E \leq -2.07$ V, a new low-frequency arc forms, indicating the existence of the mixed kinetic charge transfer and adsorption step-limited processes (Figure 8b). Parallel to the formation of the arcs in the low-frequency range, the Nyquist plots became noisier, indicating the formation of gaseous substances and/or an unstable dielectric film at the C(Mo$_2$C) electrode surface [40,46,47].

Analysis of the Nyquist plots shows that at $E \leq -1.87$ V the series resistance (R_s, estimated from the Nyquist plot data) starts to increase (Figure 9a). The increase in R_s is a clear indication of the formation of the EMIm–EMIm dimer dielectric film at the C(Mo$_2$C) electrode surface. It is interesting to note that the resistance of the mass transport process in the electrode micro-mesoporous matrix (i.e., the width of the high-frequency semicircle, R_{HFS}, calculated from the Nyquist plot data) remained almost stable in the potential range from -2.77 to -0.27 V (Figure 9b). The series capacitance C_s ($C_s = -(Z'' \, 2\pi\nu)^{-1}$, where ν is the modulation frequency in Hz), calculated at 0.1 Hz, is stable until $E = -1.77$ V (Figure 9c). A small increase in the C_s values then takes place parallel to the start of the electrochemical reduction of the EMIm$^+$ cations. C_s is maximal at $E = -1.97$ V, after which a small decrease in the C_s values takes place (Figure 9c). At $E \leq -2.67$ V, a very steep increase in C_s values was observed, indicating that quick faradic processes take place at the C(Mo$_2$C) surface (Figure 9c).

The parallel capacitance C_p values ($C_p = -Z'' \, (|Z|^2 \, 2\pi\nu)^{-1}$, where $|Z|$ is impedance modulus), calculated at 0.1 Hz, decrease in a monotonous way during the increase in the negative polarization in the 5 wt % EMImBr solution in the EMImBF$_4$–C(Mo$_2$C) interface (Figure 9d). Some stabilization in the C_p values (formation of the dielectric layer with low dielectric constant value) can be observed within the range -2.07 V $< E < -1.97$ V and at $E \leq -2.57$ V (parallel to the increase in the C_s values).

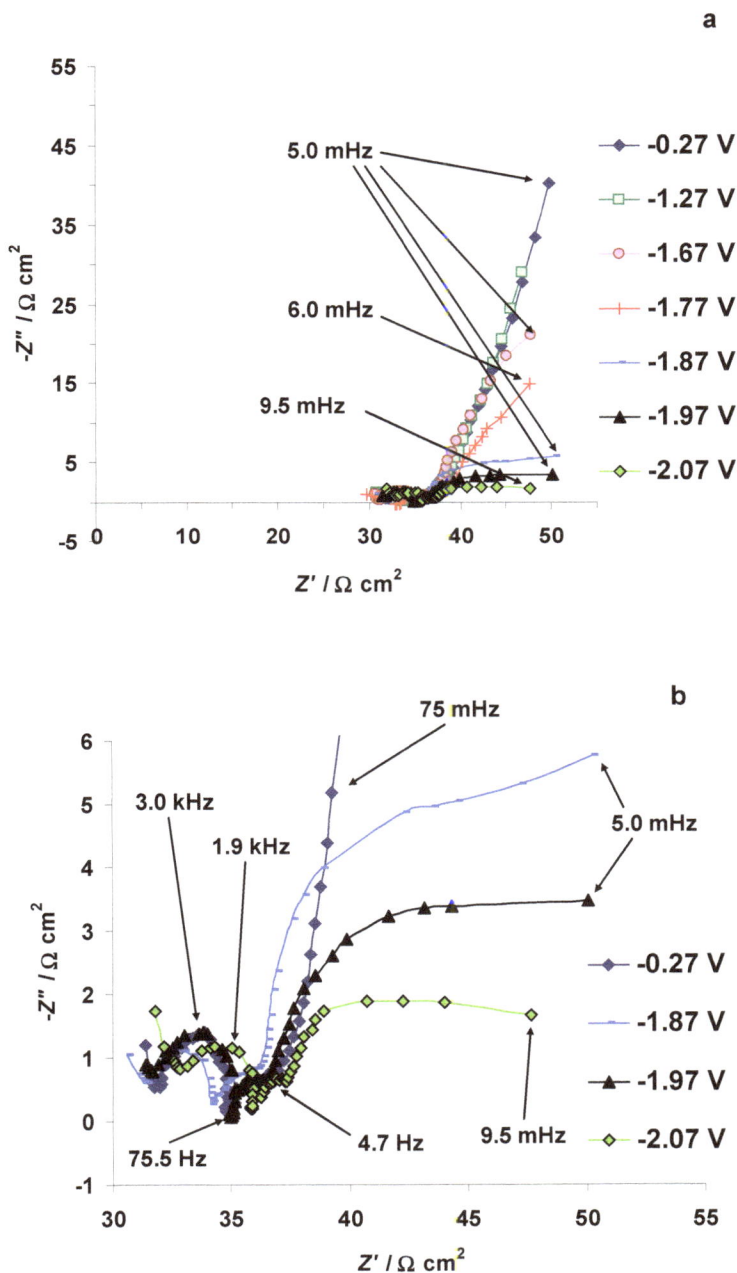

Figure 8. Electrochemical impedance spectroscopy Nyquist plots measured for the 5 wt % EMImBr solution in the EMImBF$_4$–C(Mo$_2$C) system (**a**) at variable C(Mo$_2$C) electrode negative potentials and (**b**) at selected C(Mo$_2$C) electrode negative potentials where intensive EMIm–EMIm dimer formation has started. Z' and Z'' mark the real and imaginary parts of the electrochemical impedance, respectively.

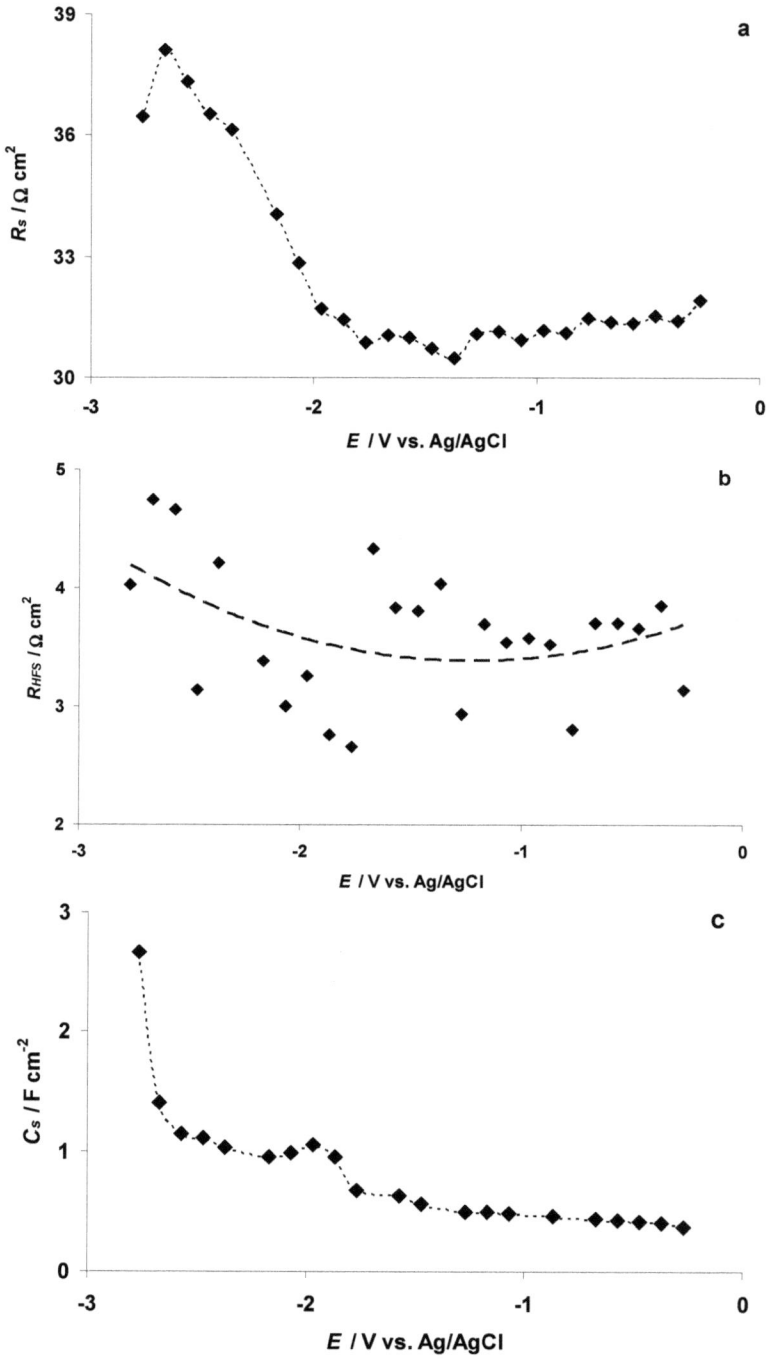

Figure 9. *Cont.*

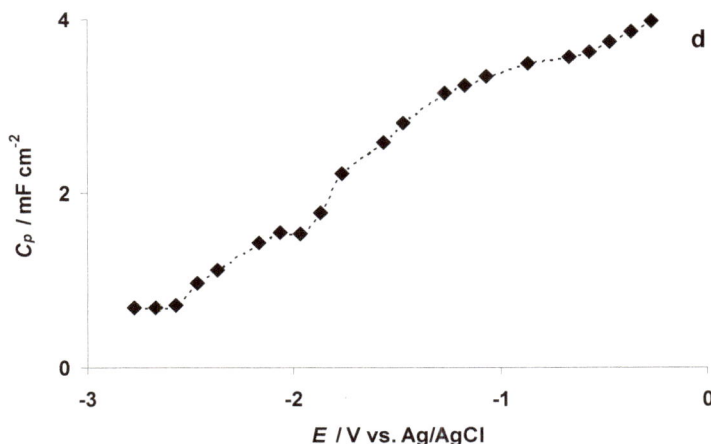

Figure 9. Illustrative data obtained from electrochemical impedance spectroscopy measurements: (**a**) series resistance (R_s); (**b**) high frequency semicircle resistance (R_{HFS}); (**c**) series capacitance (C_s) (calculated at EIS modulation frequency (v) v = 0.1 Hz); (**d**) parallel capacitance (C_p) (calculated at v = 0.1 Hz) for different consecutively measured electrochemical impedance spectra at various negative potentials for C(Mo$_2$C) in the 5 wt % EMImBr solution in EMImBF$_4$.

3.3. Characteristic Changes in the C 1s, N 1s, B 1s, F 1s, and Br 3d In Situ X-ray Spectra Collected at the Positively Polarized C(Mo$_2$C) Electrode

The X-ray photoelectron spectra for C 1s, N 1s, B 1s, F 1s, and Br 3d PE-s at specific characteristic positive potentials were recorded within the potential range from −0.27 to 1.23 V and are presented in Figure 10a–e, respectively. Positions of the PE peaks, shown in Figure 10a–e, are presented in Table S3. The PE spectra indicate a change for the aliphatic carbon (C$_5$) C 1s content at E = 0.73 V (Figure 10a), and the aliphatic carbon (C$_5$) XPS signal ratio increased from its normal value X$_{C5}$ ≈ 17% (Figure 11). The ratio of the C$_5$ XPS signal increased significantly at E = 0.93 V, obtaining a peak value X$_{C5}$ = 40% at E = 1.03 V (Figure 11). Thereafter, the ratio of C$_5$ XPS signal stabilizes at X$_{C5}$ = 31% at E ≥ 1.13 V. The in situ XPS measurements were stopped at E = 1.23 V due to the complete loss of the initial Br 3d$_{5/2}$ signal, corresponding to the final electrooxidation and disappearance of Br$^-$ ions (EMImBr) at the C(Mo$_2$C) surface (discussed hereafter).

Br 3d$_{5/2}$ and Br 3d$_{3/2}$ PE peaks are notable at the positions BE = 67.3 eV and BE = 68.3 eV, respectively (at E = 0.03 V) (Figure 10e). The shape of the XPS spectrum started to change at E ≥ 0.63 V, when new Br 3d$_{5/2}$ and Br 3d$_{3/2}$ PE peaks formed at BE = 69.6(5) eV and BE = 70.5(5) eV, respectively (Figure 10e). Parallel to the expansion of new Br 3d PE peaks, the intensity of the initial Br 3d PE peaks (marked as Br1 in Figure 11), corresponding to the Br$^-$ anion in the EMImBr, decreased (Figures 10e and 11). At E ≥ 0.93 V, the initial Br 3d$_{5/2}$ and Br 3d$_{3/2}$ PE peaks disappeared and new ones appeared, originating (very probably) from the Br$_3^-$ complex anion signal.

The recorded in situ XPS data indicate that the N 1s, B 1s, and F 1s PE signals were stable throughout the entire positive potentials region investigated (Figures 10b–d and 11). Thus, these elements have not been involved in the electrochemical oxidation reactions.

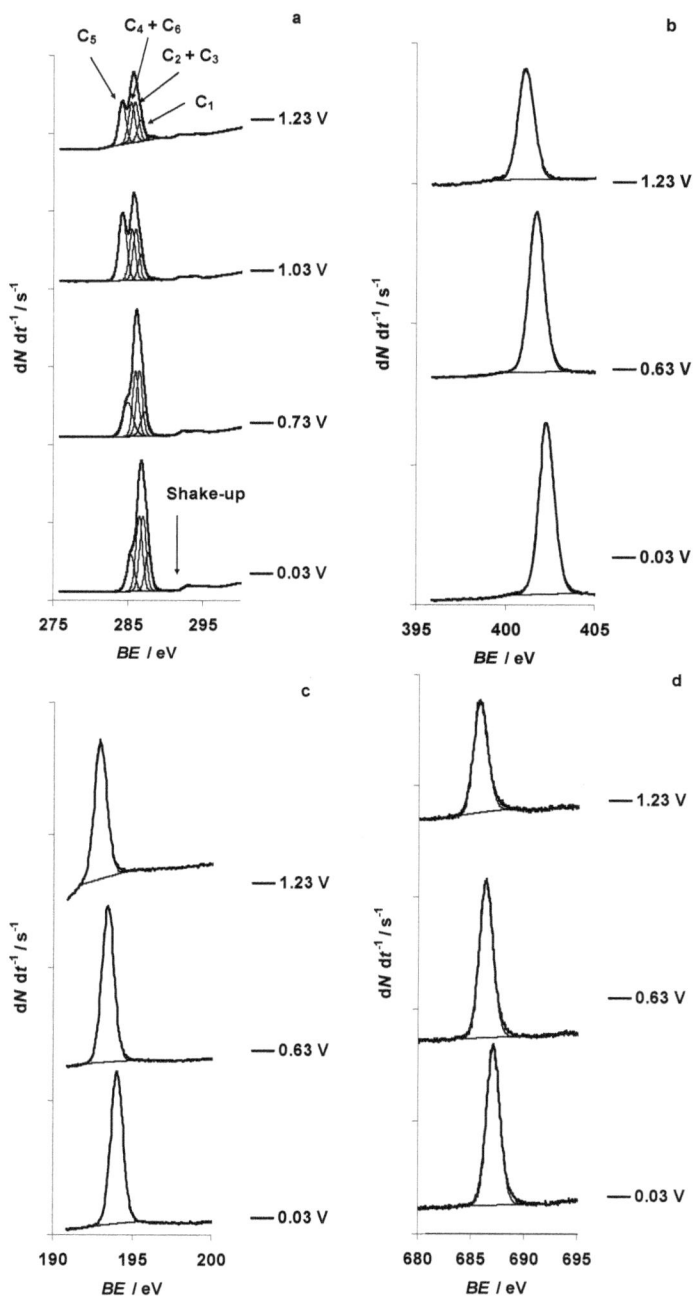

Figure 10. *Cont.*

e

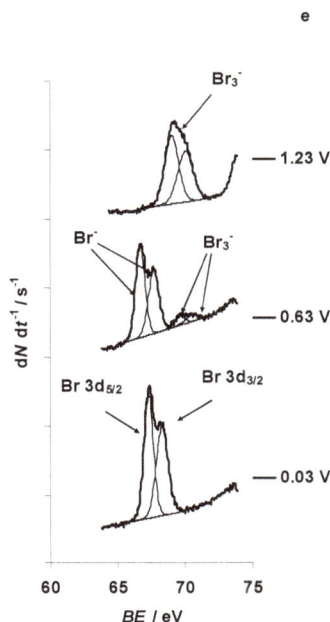

Figure 10. C 1s, N 1s, B 1s, F 1s, and Br 3d X-ray photoelectron (PE) spectra for the 5 wt % EMImBr solution in the EMImBF$_4$ mixture measured at the fixed positive potentials of the C(Mo$_2$C) electrode, noted in the figure. The binding energy (*BE*) scales for the XPS experiments have been referenced to the *BE* of the C 1s photoemission line related to aliphatic carbon (*BE* = 285.3 eV) measured for the non-polarized and grounded electrode: (**a**) C 1s PE spectra, fitted applying the four C 1s photoelectron peak model (regular lines); excitation energy was 400 eV, and the PE signal intensity (dN dt^{-1}) scale between tick marks was 500 counts s^{-1}; (**b**) N 1s PE spectra; excitation energy was 500 eV, and the PE signal intensity scale between tick marks was 200 counts s^{-1}; (**c**) B 1s PE spectra; excitation energy was 250 eV, and the PE signal intensity scale between tick marks was 200 counts s^{-1}; (**d**) F 1s PE spectra; excitation energy was 800 eV, and the PE signal intensity scale between tick marks was 10 counts s^{-1}; (**e**) Br 3d PE spectra; excitation energy was 120 eV, and the PE signal intensity scale between tick marks was 50 counts s^{-1}.

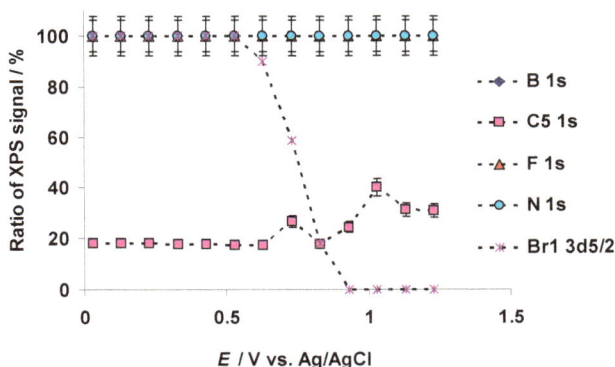

Figure 11. Dependence of EMIm$^+$ "aliphatic" carbon C 1s (marked in the figure as C5), imidazolium nitrogen atoms N 1s, BF$_4^-$ anion boron B 1s, BF$_4^-$ anion fluorine F 1s, and Br$^-$ anion bromine Br 3d$_{5/2}$ (marked in the figure as Br1 3d5/2) photoelectron peaks ratios for the 5 wt % EMImBr solution in the EMImBF$_4$ mixture at various positive potentials of the C(Mo$_2$C) electrode.

3.4. The Electrochemical Measurements Data Collected at the Positively Polarized C(Mo₂C) Electrode

In order to understand the electrochemical behavior of the 5 wt % EMImBr solution in the EMImBF$_4$–C(Mo$_2$C) interface at positive potentials, CV measurements (with the potential sweep rate 1.0 mV s^{-1}) were conducted at $p \approx 1 \times 10^{-7}$ mbar after the end of the in situ XPS measurements within the potential range from −0.27 to 1.23 V and vice versa (Figure 12a, the second CV cycle is shown). The data show an almost exponential increase in the anodic current in the forward scan at $E = 0.54$ V due to the start of the Br$^-$ anion electrooxidation to the Br$_3$$^-$ complex anion. However, it is not possible to observe the electrooxidation current peak, as the voltammogram became very noisy at $E \geq 0.74$ V (Figure 12a) caused by the intensive electrooxidation of the Br$^-$ anion and some increase in the intensity of the aliphatic carbon C$_5$ 1s XPS PE peak signal (Figure 11). The formation of the Br$_3$$^-$ complex anion could not cause this kind of electrochemical noise, and we propose that the Br$_3$$^-$ complex anion was unstable in the high vacuum condition, dissociating to Br$_2$ and Br$^-$ anions. The formation of the Br$_2$ gas at $E \geq 0.73$ V could explain the increase in the XPS vacuum chamber pressure notable in Figure 5a. (However, the sudden reduction of the XPS chamber pressure at $E > 1.13$ V (Figure 5a), could be explained by the remarkable decrease in the Br$^-$ anion concentration in the very thin electrolyte layer exposed to the exciting X-ray beam (Figure 11)) While Br$_2$ has relatively high vapor pressure (at $T \approx 22$ °C), the movement of Br$_2$ gas bubbles from the inner part of the micro-mesoporous C(Mo$_2$C) electrode to its surface and their collapse at the WE surface could explain the electrochemical noise in the CVs (Figure 12a). It should be noted that this kind of noise was not observed in the CVs recorded at $p \approx 1$ bar, where the gas evolution from the electrode surface (i.e., "bubbling") was less intense.

The CVs measured for the micro-mesoporous C(Mo$_2$C) electrode in the 5 wt % EMImBr solution in EMImBF$_4$ within the potential sweep range from 0.00 to 2.00 V (black line) and vice versa and from 0.00 to 3.00 V (gray line) and vice versa, as shown in Figure 12b. A first current maximum forms at $E = 0.86$ V, corresponding to the electrooxidation of the Br$^-$ anion to the Br$_3$$^-$ complex (indicated as E1 in Figure 12b). At ca. $E = 1.42$ V, a wide voltammetric wave was observed (marked as E2 in Figure 12b). It is possible that sharp electrooxidation current peaks are not notable due to the large energetical inhomogenity of the micro-mesoporous C(Mo$_2$C) electrode surface. Voltammetric waves of reduction processes, indicated in Figure 12b at $E = 0.80$ V as E3 and at $E = 0.37$ V as E4, were found.

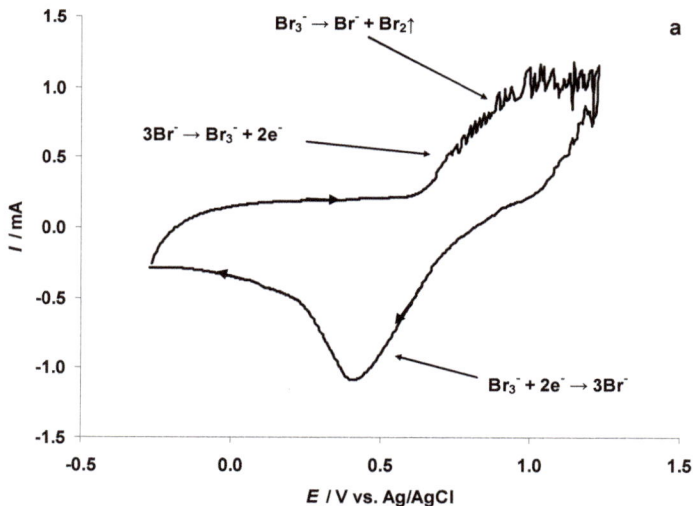

Figure 12. *Cont.*

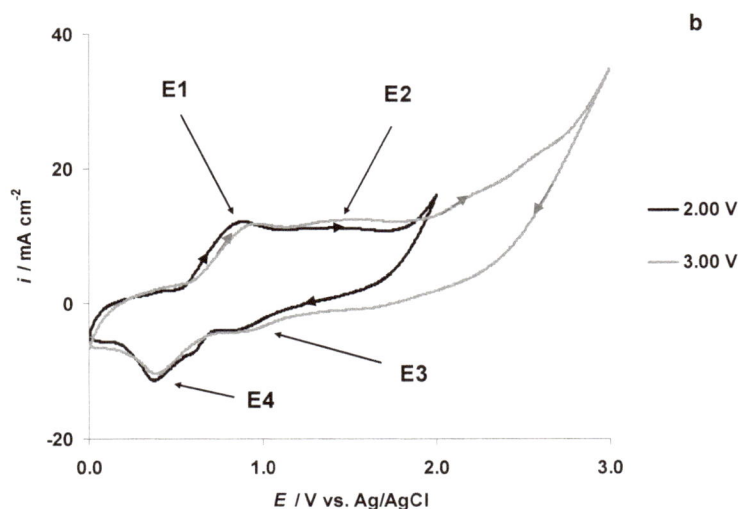

Figure 12. Cyclic voltammetry (CV) data for positively polarized C(Mo$_2$C) electrodes soaked in 5 wt % EMImBr solution in the EMImBF$_4$ solution: (**a**) located inside the XPS vacuum chamber (ca. $p = 10^{-7}$ mbar) and (**b**) inside the very dry and oxygen free Ar-filled glovebox (ca. $p = 1$ bar) within the potential sweep ranges from 0.00 to 2.00 V and vice versa (black line) and from 0.00 to 3.00 V and vice versa (gray line). Data of second cycles have been presented measured at the potential scan rate of 1.0 mV s^{-1}.

It should be noted that the separation between the $3Br^- \rightarrow Br_3^- + 2e^-$ and $2Br_3^- \rightarrow 3Br_2 + 2e^-$ processes depends significantly on the electrolyte solution used [51]. The study of Allen et al. [51] showed that the stability of the Br$_3^-$ complex in 1-buthyl-3-methylimidazolium bis(trifluoromethylsulfonyl)imide (BMIm(NTf$_2$)) was ca. 3000 times lower than that in the acetonitrile solution. The electrochemical oxidation of the Br$^-$ anion to the Br$_3^-$ complex at the platinum electrode in BMIm(NTf$_2$) was much slower (and irreversible) than at the platinum electrode soaked in an acetonitrile electrolyte [51]. On the other hand, Bennett et al. [52] demonstrated good separation, i.e., a ca. 0.5 V difference, between two consequent electrooxidation processes—$3Br^- \rightarrow Br_3^- + 2e^-$ and 2) $2Br_3^- \rightarrow 3Br_2 + 2e^-$—that take place at the glassy carbon electrode soaked in a 10 mM tetraethylammonium bromide solution in nitrobenzene.

Extending the CV sweep range up to 3.00 V (gray line, Figure 12b), the current density started to slowly increase at $E > 1.95$ V and new, low-intensity waves appeared at ca. $E = 2.25$ V and $E = 2.60$ V. However, due to the very low rate of the $2Br_3^- \rightarrow 3Br_2 + 2e^-$ reaction at the glassy carbon electrode, the exact start of this electrooxidation reaction can not be established.

Parallel with the CV measurements, EIS measurements in the potentiostatic regime (from $E = 0.03$ V up to $E = 2.03$ V) and within the frequency range from 300 kHz to 0.95 mHz were performed (Figure 13a). The $-Z''$ vs. Z' plots overlap within the potential range from $E = 0.03$ to $E = 0.53$ V. The plot measured at $E = 0.58$ V has the same shape as the previous one, measured at $E = 0.53$ V. However, the Z'' value, measured at $E = 0.58$ V and ac frequency $\nu = 0.95$ mHz, increased to -59.4 Ω cm^2, compared to the Z'' value of -75.9 Ω cm^2, obtained at $E = 0.53$ V and $\nu = 0.95$ mHz. This could be read as an early indication of the start of the electrooxidation of the Br$^-$ anion (Figures 11 and 12a,b). Increasing the 5 wt % EMImBr solution in the EMImBF$_4$–C(Mo$_2$C) interface potential, the $-Z''$ vs. Z' plot in the low frequency range preserves up to $E = 1.73$ V (Figure 13a). It is interesting to note that the high frequency semicircles are present throughout the potential range studied (0.03 V < E < 2.03 V), indicating that the micro-mesoporous structure of the C(Mo$_2$C) electrode has not been blocked with the Br$^-$ anion electrooxidation products (Figure 13b).

The Nyquist plot, measured at E = 2.63 V, contains high- and mid-frequency semicircles, and a low-frequency arc that indicate the intensification of the electrochemical oxidation processes in the 5 wt % EMImBr solution in the EMImBF$_4$–C(Mo$_2$C) interface (Figure 13c). The formation of an additional mid-frequency semicircle and a low-frequency arc corresponded to the very low-intensity wave at E = 2.60 V in CV in Figure 12b. The high-frequency semicircle disappeared and a low-frequency semicircle formed in the Nyquist plot measured at E = 2.73 V (Figure 13c). This indicates the complete blockage of the micro-mesoporous and slow charge transfer at the C(Mo$_2$C) electrode surface.

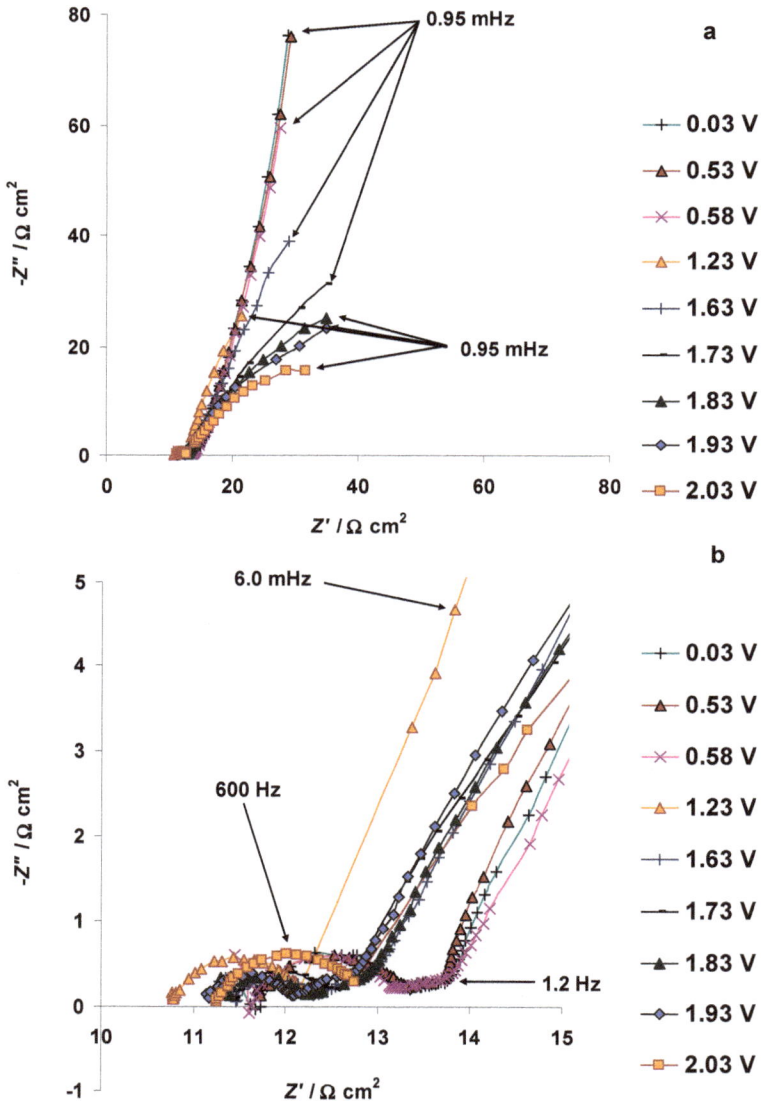

Figure 13. *Cont.*

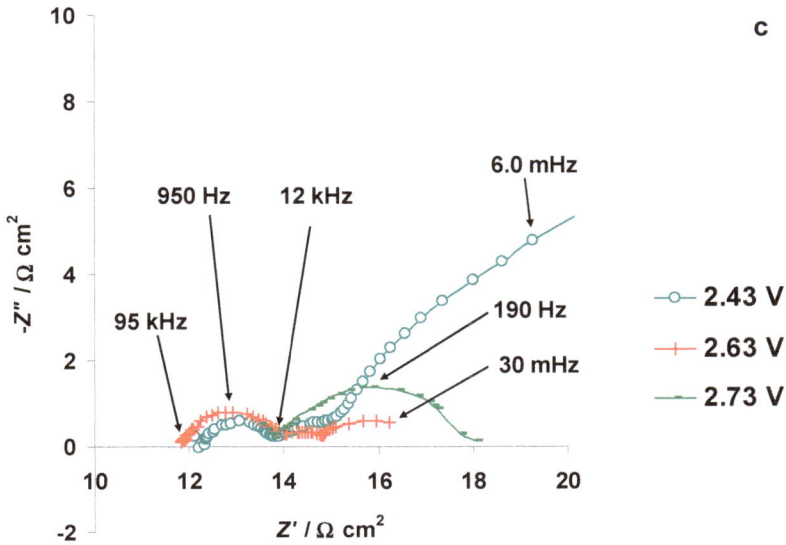

Figure 13. Electrochemical impedance spectroscopy Nyquist plots measured for the 5 wt % EMImBr solution in the EMImBF$_4$–C(Mo$_2$C) system at variable positive potentials: (**a**) from 0.03 to 2.03 V; (**b**) in the same potential region, but with the high frequency part extended; (**c**) at selected higher potentials, noted in the figure. Z' and Z'' mark the real and imaginary parts of the electrochemical impedance, respectively.

The R_s values for the 5 wt % EMImBr solution in the EMImBF$_4$–C(Mo$_2$C) system estimated from the Nyquist plot data were stable (ca. $R_s = 12 \ \Omega \ cm^2$), within the potential range 0.03 V < E < 0.73 V (Figure 14a). At $E = 0.83$ V, R_s started to decrease, parallel to the very intensive electrooxidation of the Br$^-$ anion, and a minimum value (ca. 10 Ω cm^2) at $E = 1.13$ V (E1 in Figure 12b) was observed.

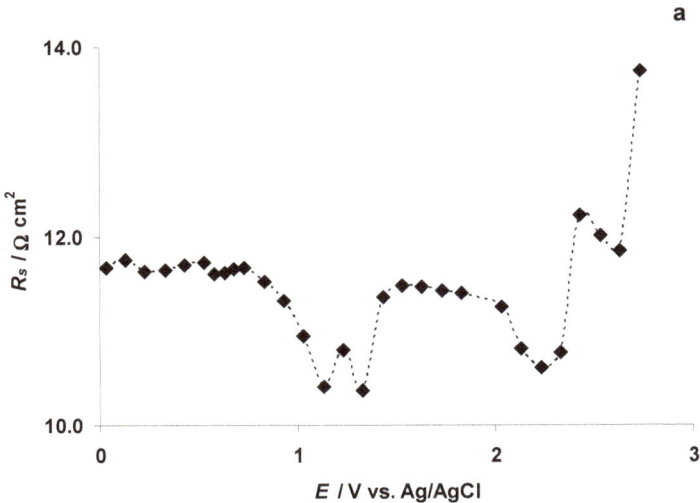

Figure 14. *Cont.*

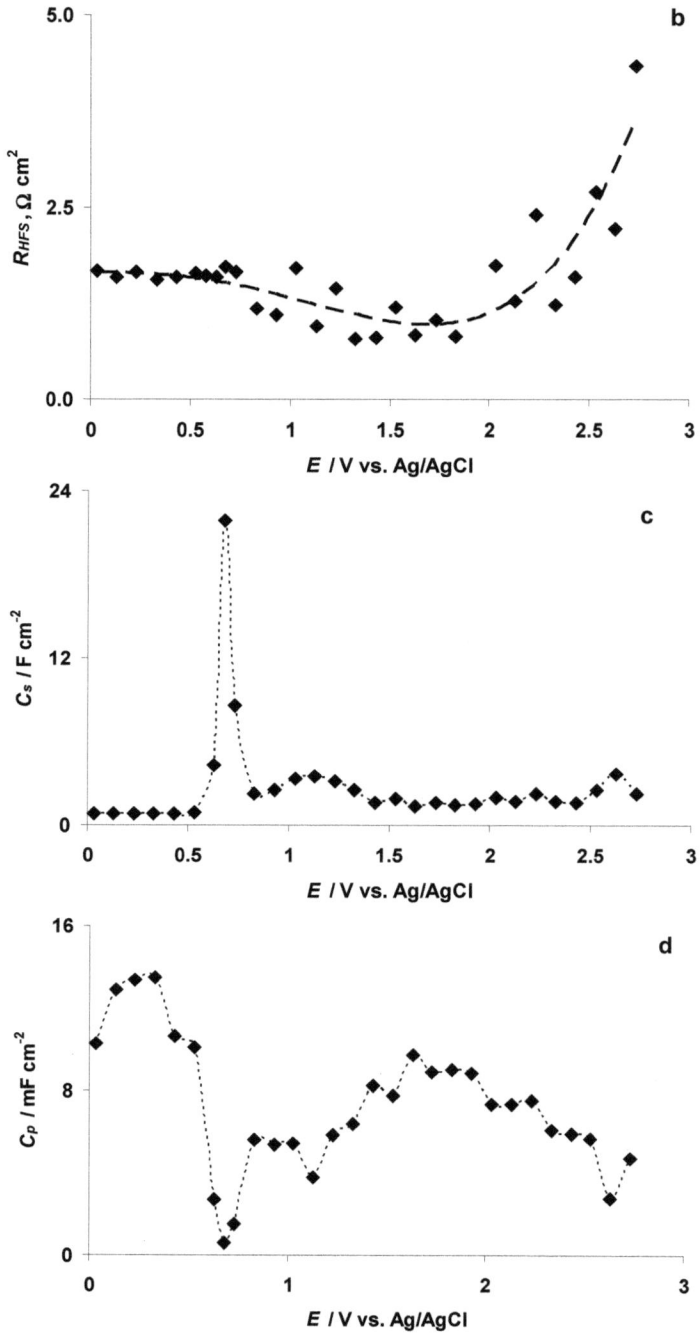

Figure 14. *Cont.*

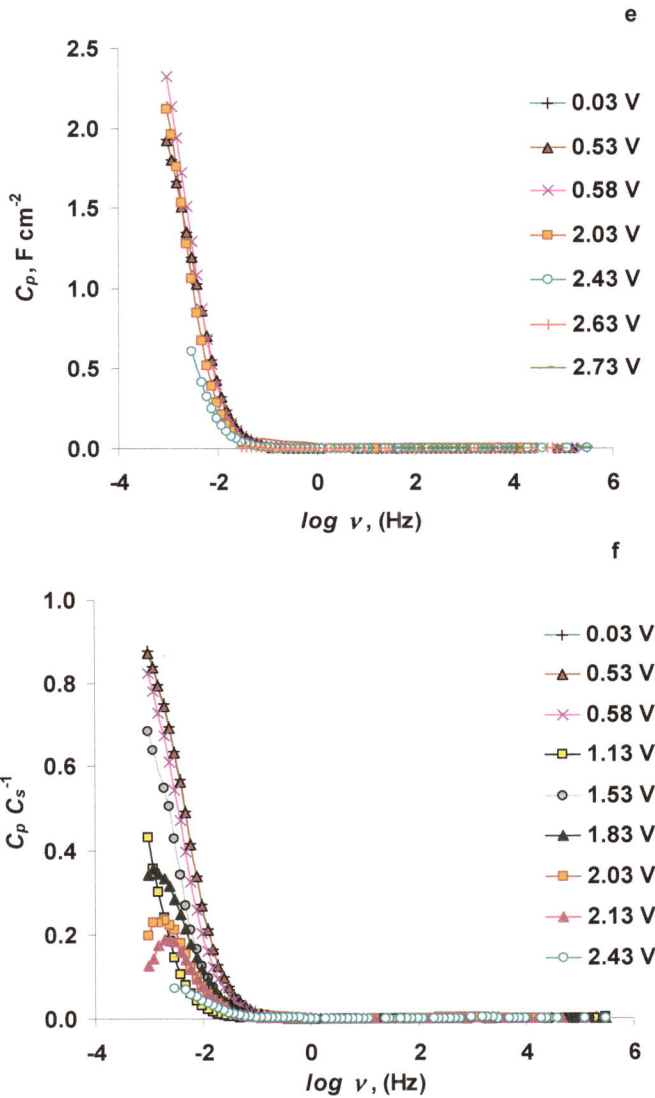

Figure 14. Illustrative data obtained from electrochemical impedance spectroscopy measurements: (**a**) —series resistance (R_s); (**b**) high frequency semicircle resistance (R_{HFS}); (**c**) series capacitance (C_s) (calculated at EIS modulation frequency (ν) $\nu = 100$ mHz); (**d**) parallel capacitance (C_p) (calculated at $\nu = 100$ mHz) for different consecutively measured electrochemical impedance spectra at various 5 wt % EMImBr solutions in the EMImBF$_4$–C(Mo$_2$C) system (i.e., C(Mo$_2$C) electrode) positive potentials; (**e**) parallel capacitance (C_p) vs. log ν (ν marks the modulation frequency in Hz) relationship for various 5 wt % EMImBr solutions in the EMImBF$_4$–C(Mo$_2$C) system positive potentials; (**f**) C_p C_s^{-1} vs. log ν (ν marks the modulation frequency in Hz) relationship for various 5 wt % EMImBr solutions in the EMImBF$_4$–C(Mo$_2$C) system at positive potentials.

The R_{HFS} (i.e., the mass transport resistance in the micro-mesoporous C(Mo$_2$C) electrode pores) values are in agreement with the R_s values at $E < 0.83$ V (Figure 14b). However, the R$_{HFS}$ values became unstable at $E > 0.73$ V if the intensive electrooxidation of the Br^- anion to the Br_3^- complex anion and Br_2 (Figures 11 and 12a,b) was observed. The increase in the R_{HFS} at $E > 1.8$ V indicates a more restricted mass transport in the C(Mo$_2$C) electrode pores.

C_s values, calculated at $\nu = 0.1$ Hz from the EIS measurements, show an intensive C_s peak ($C_s \approx 22$ F cm^{-2}) at $E = 0.68$ V (Figure 14c). The potential of the C_s peak overlaps with the maximum rate of the Br^- anion electrooxidation to the Br_3^- complex anion at micro-mesoporous C(Mo$_2$C) (defined as E1 in Figure 12b and the C microelectrode (Figure 14c,d)). The C_p values, calculated at $\nu = 0.1$ Hz from the EIS measurements, are minimal at the same potentials, where C_s has the maximum value, and intensive charge transfer processes, probably giving dielectric adsorbing products, take place (Figure 14d). C_p vs. log ν data show that the C_p values expand monotonously at log $\nu < -1.5$ (Hz) (Figure 14e). The shape of the C_p vs. log ν curves indicates the existence of a slow adsorption process inside the micro-mesoporous C(Mo$_2$C) electrode.

However, at higher C(Mo$_2$C) potentials ($E > 1.33$ V) C_p values (measured at $\nu = 0.95$ mHz) decrease (Figure 15a), but C_s values increase slightly (Figure 15b), indicating the intensification of the electrooxidation processes in the 5 wt % EMImBr solution in the EMImBF$_4$–C(Mo$_2$C) interface. At $E \geq 2.43$ V, the Nyquist plots became unstable, so the calculation of C_p was impossible.

The ratio $C_p C_s^{-1} = 0.9$ (Figure 14f), calculated at $E = 0.03$ V ($\nu = 0.95$ mHz), deviates from the value 1.0. The value 1.0 marks the ideal adsorption-limited process. Increasing the 5 wt % EMImBr solution in the EMImBF$_4$–C(Mo$_2$C) interface potential, the $C_p C_s^{-1}$ value decreased remarkably, indicating the existence of some very slow charge transfer reaction(s) at the electrode surface. At $E = 1.83$ V, a peak formed at $\nu = 1.2$ mHz in the $C_p C_s^{-1}$ vs. log ν plot ($C_p C_s^{-1} = 0.4$ at maximum peak. Continuing to increase the C(Mo$_2$C) electrode potential toward more positive values, the value of $C_p C_s^{-1}$ and the maximum of the $C_p C_s^{-1}$ vs. log ν curve moved toward higher frequency values ($C_p C_s^{-1} = 0.2$, at $E = 2.03$ V and $\nu = 1.9$ mHz).

The phase angle vs. E plot (Figure S1), obtained at $\nu = 0.1$ Hz, had a similar shape as C_p vs. E and $C_p C_s^{-1}$ vs. E plots (Figure 14e,f). It is notable that at $\nu = 0.1$ Hz the phase angle values at all positive potentials were very low. The phase angle vs. E plot (Figure S2), obtained at $\nu = 0.95$ mHz, had a shape similar to the $C_p C_s^{-1}$ vs. E relationship (Figure S3). It should be noted that the phase angle values, measured at $\nu = 0.95$ mHz, have much more negative values than those obtained at $\nu = 0.1$ Hz (Figure S1). Increasing the 5 wt % EMImBr solution in the EMImBF$_4$–C(Mo$_2$C) potential toward more positive values, the maximum phase angle value of $-71.0°$ was recorded at $E = 0.23$ V (Figure S2).

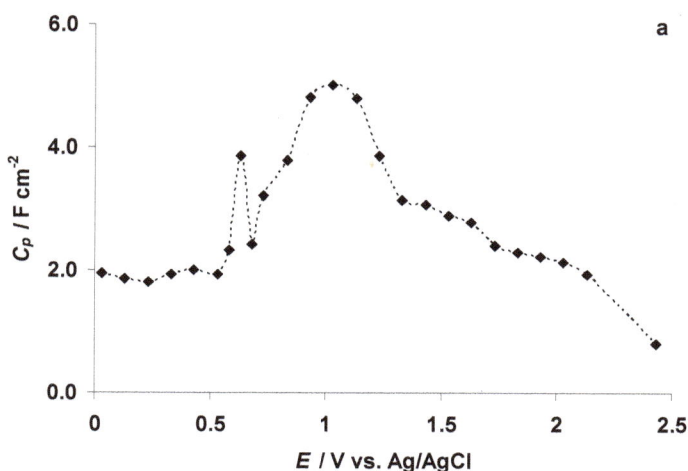

Figure 15. *Cont.*

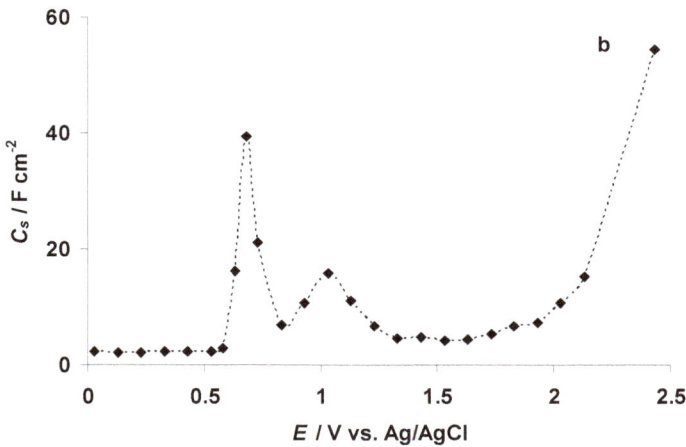

Figure 15. Data obtained from electrochemical impedance spectroscopy measurements: (**a**) parallel capacitance (C_p) and (**b**) series capacitance (C_s) values (calculated at $\nu = 0.95$ mHz) for the 5 wt % EMImBr solution in the EMImBF$_4$–C(Mo$_2$C) system at various positive potentials.

The log $|Z''|$ vs. log ν data (Figure 16) indicate that the linear relationship exists only at very low frequencies ($\nu < 0.30$ Hz). The slope and the length in the linear part of these plots depend somewhat on the potential applied. The slopes of the linear parts of the log $|Z''|$ vs. log ν data are in the range from -0.8 to -0.7 (at 0.03 V $< E < 2.33$ V), indicating that mixed kinetic oxidation/adsorption processes prevail at the micro-mesoporous C(Mo$_2$C) electrode within this potential region.

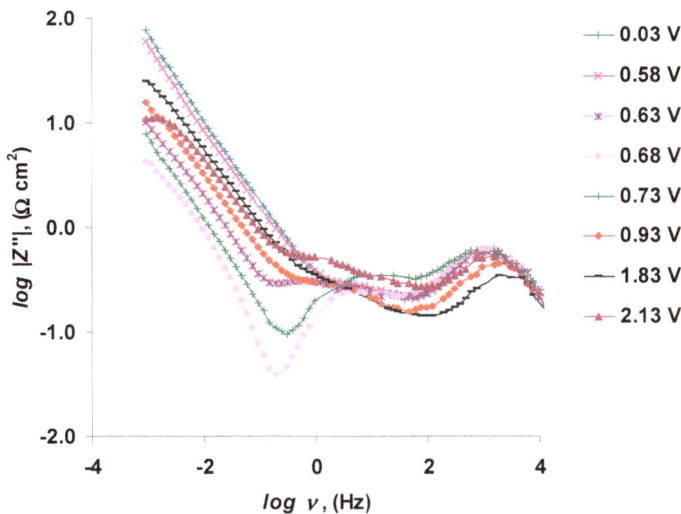

Figure 16. Electrochemical impedance spectroscopy (EIS) data: log of the imaginary part of impedance (Z'') vs. log of EIS modulation frequency (ν) dependences for the 5 wt % EMImBr solution in the EMImBF$_4$–C(Mo$_2$C) interface at various positive potentials, noted in the figure.

4. Conclusions

The in situ X-ray photoelectron spectroscopy (XPS) data for aliphatic carbon (C$_5$) C$_5$ 1s, N 1s, B 1s, F 1s, and Br 3d$_{5/2}$ were measured for a 5 wt % 1-ethyl-3-methylimidazolium bromide solution in the 1-ethyl-3-methylimidazolium tetrafluoroborate–molybdenum carbide-derived carbon electrode interface at a residual water (210 ppm) level. The calculated data indicated that the 1s electrons binding energy vs. potential ($dBE\ dE^{-1}$) plots for C, N, B, and F elements were all linear with the slope $dBE\ dE^{-1} = -1$ eV V^{-1} within the potential range from -1.17 to 1.23 V (i.e., in the region of ideal polarization). At more negative potentials (-2.07 V $< E < -1.17$ V), the $dBE\ dE^{-1}$ value was nearly -0.5 eV V^{-1} for C_5 1s, N1 1s, B1 1s, F 1s, and Br 3d$_{5/2}$ PEs. It was established that the reduction of the $dBE\ dE^{-1}$ slope's absolute value, twice at $E \leq -1.17$ V, was connected with the start of the formation of gas bubbles at the $C(Mo_2C)$ electrode. The formation of a new B 1s PE peak, corresponding to the B–O bond, was caused by the electroreduction of the residual water adsorbed at the micro-mesoporous $C(Mo_2C)$ electrode.

The cyclic voltammetry (CV) measurements, performed in high vacuum conditions (ca. $p = 10^{-7}$ mbar), indicated that the electrooxidation of the Br$^-$ anion started at $E = 0.54$ V in the 5 wt % EMImBr solution in the EMImBF$_4$–C(Mo$_2$C)) interface. At $E \geq 0.74$ V, the measured cyclic voltammogram became very noisy, indicating the instability of the Br$_3^-$ complex under vacuum (and the evaporation of Br$_2$). The Br 3d$_{5/2}$ XPS data indicated that the intensity of the Br$^-$ anion electrooxidation at $E \geq 0.63$ V (as the arbitrary intensity of the corresponding photoelectron (PE) peak) started to reduce and even disappeared at $E \geq 0.93$ V. Parallel to the start of the decrease in the initial Br 3d$_{5/2}$ and Br 3d$_{3/2}$ PE peaks at $E = 0.63$ V, new Br 3d$_{5/2}$ and 3d$_{3/2}$ PE peaks (at ca. $\Delta BE = 3$ eV higher BEs) formed, corresponding to the formation of the Br$_3^-$ complex anion.

The CV method was not sensitive enough to separate the 3Br$^-$ $\rightarrow$ Br$_3^-$ + 2e$^-$ and 2Br$_3^-$ $\rightarrow$ 3Br$_2$ + 2e$^-$ processes taking place in the 5 wt % EMImBr solution in the EMImBF$_4$–micro-mesoporous $C(Mo_2C)$ interface. However, separation and quantitative analysis of these electrochemical reactions is possible based on Br 3d$_{5/2}$ in situ XPS and electrochemical impedance data. On the other hand, CV measurements provided useful information for 5 wt % EMImBr solutions in an EMImBF$_4$–carbon fiber microelectrode system due to the larger ratios of signal to noise and of faradic current to charging current.

Supplementary Materials: The following are available online at http://www.mdpi.com/2079-4991/9/2/304/s1. Table S1: C 1s, N 1s, B 1s, F 1s, and Br 3d PE binding energies (eV) measured by in situ XPS for the 5 wt % EMImBr solution in the EMImBF$_4$–C(Mo$_2$C) system polarized at various negative potentials (corresponding spectra are shown in Figure 2a–e). Table S2: dBE vs. dE (eV V^{-1}) slopes for aliphatic carbon (C_5) C 1s, initial nitrogen (N1) N 1s, initial boron (B1) B 1s, F 1s, and initial bromine (Br) Br 3d$_{5/2}$ photoelectron binding energies measured for the 5 wt % EMImBr solution in the EMImBF$_4$–C(Mo$_2$C) system for various potential ranges. Table S3: C 1s, N 1s, B 1s, F 1s, and Br 3d PE binding energies (eV) measured by in situ XPS for the 5 wt % EMImBr solution in the EMImBF$_4$–C(Mo$_2$C) system polarized at various positive potentials (corresponding spectra are shown in Figure 10a–e). Figure S1: Electrochemical impedance spectroscopy phase angle data for the 5 wt % EMImBr solution in EMImBF$_4$, measured at $\nu = 0.1$ Hz and at various C(Mo$_2$C) electrode positive potentials. Figure S2: Electrochemical impedance spectroscopy phase angle data for the 5 wt % EMImBr solution in EMImBF$_4$, measured at $\nu = 0.95$ mHz and at various C(Mo$_2$C) electrode positive potentials. Figure S3: The parallel capacitance (C_p) and series capacitance (C_s) ratio ($C_p\ C_s^{-1}$) data at various C(Mo$_2$C) electrode positive potentials and $\nu = 0.95$ mHz, obtained for the 5 wt % EMImBr solution in the EMImBF$_4$–C(Mo$_2$C) interface.

Author Contributions: J.K. wrote the manuscript, conducted all electrochemical measurements, and performed the in situ XPS and electrochemical measurements data analysis; A.T. constructed the measurement cell and performed the in situ XPS measurements; R.P. and E.N. performed the in situ XPS measurements; E.L. has reviewed and approved the manuscript.

Funding: This research was funded by the Estonian Research Council (projects IUT20-13 and IUT2-25) and the European Regional Development Fund (Estonian Centre of Excellence (1.01.2016–1.03.2023)).

Acknowledgments: We are also grateful to the staff of Max-Lab (Lund University, Sweden) for the assistance and co-operation during the measurements, to Martin Vilbaste (Chair of Analytical Chemistry, Institute of Chemistry, University of Tartu) for the analysis of the water concentration in the 5 wt % EMImBr solution in the EMImBF$_4$ studied, to Indrek Tallo for the synthesis of the C(Mo$_2$C) electrode, and to Tavo Romann for coating the C(Mo$_2$C) electrode with a thin Al film.

Conflicts of Interest: The authors declare no conflict of interest.

References

1. Wang, G.; Zhang, L.; Zhang, J. A review of electrode materials for electrochemical supercapacitors. *Chem. Soc. Rev.* **2012**, *41*, 797–828. [CrossRef] [PubMed]
2. Zhang, L.; Zhao, X.S. Carbon-based materials as supercapacitor electrodes. *Chem. Soc. Rev.* **2009**, *38*, 2520–2531. [CrossRef] [PubMed]
3. Beguin, F.; Presser, V.; Balducci, A.; Frackowiak, E. Carbons and electrolytes for Advanced Supercapacitors. *Adv. Mater.* **2014**, *26*, 2219–2251. [CrossRef] [PubMed]
4. Theerthagiri, J.; Karuppasamy, K.; Durai, G.; Rana, A.H.S.; Arunachalam, P.; Sangeetha, K.; Kuppusami, P.; Kim, H.-S. Recent Advances in Metal Chalcogenides (MX.; X = S, Se) Nanostructures for Electrochemical Supercapacitor Applications: A Brief Review. *Nanomaterials* **2018**, *8*, 256. [CrossRef] [PubMed]
5. Ghosh, A.; Lee, Y.H. Carbon-based electrochemical capacitors. *Chem. Sus. Chem.* **2012**, *5*, 480–499. [CrossRef] [PubMed]
6. Miller, J.R.; Burke, A.F. Electrochemical capacitors: Challenges and opportunities for real-world applications. *Interface* **2008**, *17*, 53–57.
7. Miller, J.R.; Simon, P. Electrochemical capacitors for energy management. *Science* **2008**, *321*, 651–652. [CrossRef] [PubMed]
8. Kötz, R.; Carlen, M. Principles and applications of electrochemical capacitors. *Electrochim. Acta* **2000**, *45*, 2483–2498. [CrossRef]
9. Simon, P.; Gogotsi, Y. Materials for electrochemical capacitors. *Nature Mater.* **2008**, *7*, 845–854. [CrossRef] [PubMed]
10. Volfkovich, Y.M.; Serdyuk, T.M. Electrochemical capacitors. *Russ. J. Electrochem.* **2002**, *38*, 935–959. [CrossRef]
11. Pell, W.G.; Conway, B.E.; Adams, W.A.; de Oliveira, J. Electrochemical efficiency in multiple discharge/recharge cycling of supercapacitors in hybrid EV applications. *J. Power Sources* **1999**, *80*, 134–141. [CrossRef]
12. Jänes, A.; Eskusson, J.; Kanarbik, R.; Saar, A.; Lust, E. Surface analysis of supercapacitor electrodes after long-lasting constant current tests in organic electrolyte. *J. Electrochem. Soc.* **2012**, *159*, A1141–A1147. [CrossRef]
13. Romann, T.; Oll, O.; Pikma, P.; Tamme, H.; Lust, E. Surface chemistry of carbon electrodes in 1-ethyl-3-methylimidazolium tetrafluoroborate ionic liquid—An in situ infrared study. *Electrochim. Acta* **2014**, *125*, 183–190. [CrossRef]
14. Kurig, H.; Vestli, M.; Tõnurist, K.; Jänes, A.; Lust, E. Influence of room temperature ionic liquid anion chemical composition and electrical charge delocalization on the supercapacitor properties. *J. Electrochem. Soc.* **2012**, *159*, A944–A951. [CrossRef]
15. Jänes, A.; Thomberg, T.; Kurig, H.; Lust, E. Nanoscale fine-tuning of porosity of carbide-derived carbon prepared from molybdenum carbide. *Carbon* **2009**, *47*, 23–29. [CrossRef]
16. Jänes, A.; Permann, L.; Arulepp, M.; Lust, E. Electrochemical characteristics of nanoporous carbide-derived carbon materials in non-aqueous electrolyte solutions. *Electrochem. Commun.* **2004**, *6*, 313–318. [CrossRef]
17. Thomberg, T.; Jänes, A.; Lust, E. Energy and power performance of electrochemical double-layer capacitors based on molybdenum carbide derived carbon. *Electrochim. Acta* **2010**, *55*, 3138–3143. [CrossRef]
18. Jänes, A.; Thomberg, T.; Tõnurist, K.; Kurig, H.; Laheäär, A.; Lust, E. Micro- and mesoporous carbide-derived carbon materials and polymer membranes for supercapacitors. *ECS Transact.* **2008**, *16*, 57–67. [CrossRef]
19. Von Czarnecki, P.; Ahrens, M.; Iliev, B.; Schubert, T.J.S. Ionic liquid based electrolytes for electrical storage. *ECS Transact.* **2017**, *77*, 79–87. [CrossRef]
20. Chandra, A. Supercapacitors: An alternative technology for energy storage. *Proc. Natl. Acad. Sci. India Sect. A Phys. Sci.* **2012**, *82*, 79–90. [CrossRef]
21. Jänes, A.; Eskusson, J.; Mattisen, L.; Lust, E. Electrochemical behaviour of hybrid devices based on Na_2SO_4 and Rb_2SO_4 neutral aqueous electrolytes and carbon electrodes within wide cell potential region. *J. Solid State Electrochem.* **2015**, *19*, 769–783. [CrossRef]
22. Gao, X.; Zu, L.; Cai, X.; Li, C.; Lian, H.; Liu, Y.; Wang, X.; Cui, X. High Performance of Supercapacitor from PEDOT:PSS Electrode and Redox Iodide Ion Electrolyte. *Nanomaterials* **2018**, *8*, 335. [CrossRef] [PubMed]
23. Gao, H.; Lian, K. Proton-conducting polymer electrolytes and their applications in solid supercapacitors: A review. *RSC Adv.* **2014**, *4*, 33091–33113. [CrossRef]

24. Ehsani, A.; Heidari, A.A.; Shiri, H.M. Electrochemical pseudocapacitors based on ternary nanocomposite of conductive polymer/graphene/metal oxide: An introduction and review to it in recent studies. *Chem. Rec.* **2018**, *18*, 1–20. [CrossRef] [PubMed]

25. Thangavel, R.; Kannan, A.G.; Ponraj, R.; Thangavel, V.; Kim, D.-W.; Lee, Y.-S. High-energy green supercapacitor driven by ionic liquid electrolytes as an ultra-high stable next-generation energy storage device. *J. Power Sources* **2018**, *383*, 102–109. [CrossRef]

26. Gonzales, A.; Goikolea, E.; Barrena, J.A.; Mysyk, R. Review on supercapacitors: Technologies and materials. *Renew. Sustain. Energy Rev.* **2016**, *58*, 1189–1206. [CrossRef]

27. Meng, Q.; Cai, K.; Chen, Y.; Chen, L. Research progress on conducting polymer based supercapacitor electrode materials. *Nano Energy* **2017**, *36*, 268–285. [CrossRef]

28. Zhang, W.; Feng, P.; Chen, J.; Sun, Z.; Zhao, B. Electrically conductive hydrogels for flexible energy storage systems. *Prog. Polym. Sci.* **2019**, *88*, 220–240. [CrossRef]

29. Shi, Y.; Zhang, J.; Pan, L.; Shi, Y.; Yu, G. Energy gels: A bio-inspired material platform for advanced energy applications. *Nano Today* **2016**, *11*, 738–762. [CrossRef]

30. Väärtnõu, M.; Lust, E. Impedance characteristics of iodide ions adsorption on Bi single crystal planes in ethanol. *J. Electroanal. Chem.* **2004**, *565*, 211–218. [CrossRef]

31. Kim, K.-S.; Shin, B.-K.; Lee, H. Physical and electrochemical properties of 1-butyl-3-metylimidazolium bromide, 1-butyl-3-methylimidazolium iodide, and 1-butyl-3-metylimidazolium tetrafluoroborate. *Korean J. Chem. Eng.* **2004**, *21*, 1010. [CrossRef]

32. Siinor, L.; Siimenson, C.; Lust, K.; Lust, E. Mixture of 1-ethyl-3-methylimidazolium tetrafluoroborate and 1-ethyl-3-methylimidazolium iodide: A new potential high capacitance electrolyte for EDLCs. *Electrochem. Commun.* **2013**, *35*, 5–7. [CrossRef]

33. Siinor, L.; Poom, J.; Siimenson, C.; Lust, K.; Lust, E. Electrochemical characteristics pyrolytic graphite | mixture of 1-ethyl-3-methylimidazolium tetrafluoroborate and 1-ethyl-3-methylimidazolium iodide interface. *J. Electroanal. Chem.* **2014**, *719*, 133–137. [CrossRef]

34. Siimenson, C.; Siinor, L.; Lust, K.; Lust, E. The electrochemical characteristics of the mixture of 1-ethyl-3-methylimidazolium tetrafluoroborate and 1-ethyl-3-methylimidazolium iodide. *J. Electroanal. Chem.* **2014**, *730*, 59–64. [CrossRef]

35. Lust, E.; Siinor, L.; Kurig, H.; Romann, T.; Ivaništšev, V.; Siimenson, C.; Thomberg, T.; Kruusma, J.; Lust, K.; Pikma, P.; et al. Characteristics of Capacitors Based on Ionic Liquids: From Dielectric Polymers to Redox-Active Adsorbed Species. *ECS Transact.* **2016**, *75*, 161–170. [CrossRef]

36. Oll, O.; Siimenson, C.; Lust, K.; Gorbatovski, G.; Lust, E. Specific adsorption from an ionic liquid: Impedance study of iodide ion adsorption from a pure halide ionic liquid at bismuth single crystal planes. *Electrochim. Acta* **2017**, *247*, 910–919. [CrossRef]

37. Yamazaki, S.; Ito, T.; Yamagata, M.; Ishikawa, M. Non-aqueous electrochemical capacitor utilizing electrolytic redox reactions of bromide species in ionic liquid. *Electrochim. Acta* **2012**, *86*, 294–297. [CrossRef]

38. Siimenson, C.; Lembinen, M.; Oll, O.; Läll, L.; Tarkanovskaja, M.; Ivaništšev, V.; Siinor, L.; Thomberg, T.; Lust, K.; Lust, E. Electrochemical Investigation of 1-Ethyl-3-methylimidazolium Bromide and Tetrafluoroborate Mixture at Bi(111) Electrode Interface. *J. Electrochem. Soc.* **2016**, *163*, H723–H730. [CrossRef]

39. Gastrol, D.; Walkowiak, J.; Fic, K.; Frackowiak, E. Enhancement of the carbon electrode capacitance by brominated hydroquinones. *J. Power Sources* **2016**, *326*, 587–594. [CrossRef]

40. Kruusma, J.; Tõnisoo, A.; Pärna, R.; Nõmmiste, E.; Lust, E. Influence of the negative potential of molybdenum carbide derived carbon electrode on the in situ synchrotron radiation activated X-ray photoelectron spectra of 1-ethyl-3-methylimidazolium tetrafluoroborate. *Electrochim. Acta* **2016**, *206*, 419–426. [CrossRef]

41. Tõnisoo, A.; Kruusma, J.; Pärna, R.; Kikas, A.; Hirsimäki, M.; Nõmmiste, E.; Lust, E. In Situ XPS Studies of Electrochemically Negatively Polarized Molybdenum Carbide Derived Carbon Double Layer Capacitor Electrode. *J. Electrochem. Soc.* **2013**, *160*, A1084–A1093. [CrossRef]

42. Foelske-Schmitz, A.; Weingarth, D.; Kötz, R. Quasi in situ XPS study of electrochemical oxidation and reduction of highly oriented pyrolytic graphite in [1-ethyl-3-methylimidazolium][BF$_4$] electrolytes. *Electrochim. Acta* **2011**, *56*, 10321–10331. [CrossRef]

43. Smith, E.F.; Rutten, F.J.M.; Villar-Garcia, I.J.; Briggs, D.; Licence, P. Ionic Liquids in Vacuo: Analysis of Liquid Surfaces Using Ultra-High-Vacuum Techniques. *Langmuir* **2006**, *22*, 9386–9392. [CrossRef] [PubMed]

44. Smith, E.F.; Villar-Garcia, I.J.; Briggs, D.; Licence, P. Ionic liquids *in vacuo*; solution-phase X-ray photoelectron spectroscopy. *Chem. Commun.* **2005**, 5633–5635. [CrossRef] [PubMed]
45. NIST X-ray Photoelectron Spectroscopy Database, NIST Standard Reference Database 20, Version 4.1. Available online: http://srdata.nist.gov/xps/EnergyTypeValSrch.aspx (accessed on 14 February 2019).
46. Kruusma, J.; Tõnisoo, A.; Pärna, R.; Nõmmiste, E.; Vahtrus, M.; Siinor, L.; Tallo, I.; Romann, T.; Lust, E. Influence of Iodide Ions Concentration on the Stability of 1-Ethyl-3-methylimidazolium Tetrafluoroborate | Molybdenum Carbide Derived Carbon Electrode Interface. *J. Electrochem. Soc.* **2017**, *164*, A1110–A1119. [CrossRef]
47. Kruusma, J.; Tõnisoo, A.; Pärna, R.; Nõmmiste, E.; Kuusik, I.; Vahtrus, M.; Tallo, I.; Romann, T.; Lust, E. The Electrochemical Behavior of 1-Ethyl-3-Methyl Imidazolium Tetracyanoborate Visualized by In Situ X-ray Photoelectron Spectroscopy at the Negatively and Positively Polarized Micro-Mesoporous Carbon Electrode. *J. Electrochem. Soc.* **2017**, *164*, A3393–A3402. [CrossRef]
48. Cheek, G.T.; O'Grady, W.E.; Lawrence, S.H. Determination of Water in 1-Ethyl-3-methylimidazolium tetrafluoroborate. *ECS Transact.* **2007**, *2*, 1–5. [CrossRef]
49. Hamm, U.W.; Lazarescu, V.; Kolb, D.M. Adsorption of pyrazine on Au(111) and Ag(111) electrodes an *ex situ* XPS study. *J. Chem. Soc. Faraday Trans.* **1996**, *92*, 3785–3790. [CrossRef]
50. Zhou, W.; Kolb, D.M. Influence of an electrostatic potential at the metal/electrolyte interface on the electron binding energy of adsorbates as probed by X-ray photoelectron spectroscopy. *Surf. Sci.* **2004**, *573*, 176–182. [CrossRef]
51. Allen, G.D.; Buzzeo, M.C.; Villagrán, C.; Hardacre, C.; Compton, R.G. A mechanistic study of the electro-oxidation of bromide in acetonitrile and the room temperature ionic liquid, 1-butyl-3-methylimidazolium bis(trifluoromethylsulfonyl)imide at platinum electrodes. *J. Electroanal. Chem.* **2005**, *575*, 311–320. [CrossRef]
52. Bennett, B.; Chang, J.; Bard, A.J. Mechanism of the Br^-/Br_2 Redox Reaction on Platinum and Glassy Carbon Electrodes in Nitrobenzene by Cyclic Voltammetry. *Electrochim. Acta* **2016**, *219*, 1–9. [CrossRef]

nanomaterials

MDPI

Article

Structural and Thermal Characterisation of Nanofilms by Time-Resolved X-ray Scattering

Anton Plech [1,*], Bärbel Krause [1], Tilo Baumbach [1,2], Margarita Zakharova [3], Soizic Eon [4], Caroline Girmen [4], Gernot Buth [1] and Hartmut Bracht [4]

[1]   Institute for Photon Science and Synchrotron Radiation, Karlsruhe Institute of Technology,
     D-76021 Karlsruhe, Germany; baerbel.krause@kit.edu (B.K.); tilo.baumbach@kit.edu (T.B.);
     gernot.buth@kit.edu (G.B.)
[2]   Laboratory for Applications of Synchrotron Radiation, Karlsruhe Institute of Technology,
     D-76021 Karlsruhe, Germany
[3]   Institute of Microstructure Technology, Karlsruhe Institute of Technology, D-76021 Karlsruhe, Germany;
     margarita.zakharova@partner.kit.edu
[4]   Institute of Materials Physics, University of Muenster, D-48149 Münster, Germany;
     soizic.eon@laposte.net (S.E.); girmen@uni-muenster.de (C.G.); bracht@uni-muenster.de (H.B.)
*   Correspondence: anton.plech@kit.edu; Tel.: +49-721-608-28665

Received: 1 March 2019; Accepted: 22 March 2019; Published: 1 April 2019

Abstract: High time resolution in scattering analysis of thin films allows for determination of thermal conductivity by transient pump-probe detection of dissipation of laser-induced heating, TDXTS. We describe an approach that analyses the picosecond-resolved lattice parameter reaction of a gold transducer layer on pulsed laser heating to determine the thermal conductivity of layered structures below the transducer. A detailed modeling of the cooling kinetics by a Laplace-domain approach allows for discerning effects of conductivity and thermal interface resistance as well as basic depth information. The thermal expansion of the clamped gold film can be calibrated to absolute temperature change and effects of plastic deformation are discriminated. The method is demonstrated on two extreme examples of phononic barriers, isotopically modulated silicon multilayers with very small acoustic impedance mismatch and silicon-molybdenum multilayers, which show a high resistivity.

Keywords: thin films; multilayers; thermal conductivity; thermal expansion; laser heating; synchrotron pump-probe powder scattering

1. Introduction

Nanoscale analysis of materials profits from the high brilliance of synchrotron-based sources by e.g., real-space or reciprocal space methods. Apart from the high resolution in real and reciprocal space the time structure of light emission at synchrotrons adds the ability to investigate dynamic processes, such as lattice motion or dissipation. One particularly important aspect of many functional materials on the nanoscale are thermal properties. In a large class of applications, such as in semiconductor integrated circuits, it is important to dissipate heat efficiently to limit heat-induced damages. On the other hand, heat conduction is an unwanted dissipation process for thermoelectric devices, which reduces the efficiency of charge collection through thermopower [1,2]. In the second case, optimization of thermoelectric materials involved minimization of the phonon contribution to thermal transport [3–5]. This can happen through introduction of interfaces with acoustic impedance mismatch or defects (pointlike or particles) [6–8], which increase phonon scattering. Multilayers, in particular, represent an interesting class of phonon barriers. The periodicity of stacking perpendicular to the surface leads to a modification of phonon propagation that can be interpreted as a new Brillouin zone, leading to phonon reflection and band bending [9–11]. Consequently, the pass band for phonons may be modified further than just by introducing individual interfaces. In analogy to optical coating a stop band may be forming [12–14], but for short periods Bloch-wave like phonon transmission can set in [15,16]. The case of layered isotopic substitution drew attention due to the fact that such crystals may electronically behave as bulk, but modulate phonon transport [14,17].

Consequently, a precise evaluation of thermal conduction on a nanoscale is very important. On surfaces several approaches are being used, the 3ω method that explores the dependence of the electrical conductivity of conducting structures during alternating-current resistive heating on top of the investigated structures as well as the dependence of optical reflectivity of transducer layers on temperature (TDTR) [18–20]. Both analyze the cooling time scale of the thermal conductivity in the structure below the surface. In a similar way, cooling of nanostructures can be sensed after pulsed heating to evaluate thermal transport in bulk-like conditions [21–23]. While optical methods are readily available in laboratories, we describe the same concept of pulsed laser heating, but probing the temperature by X-ray scattering, which adds better access to nanoscale structure [24–27]. X-ray scattering can address temperature (change) by lattice parameter (expansion), at the same time giving access to phonon modes [11,28,29] or sub-surface resolution.

We explore the limits of the method of Time-domain X-ray thermal scattering (TDXTS, name given in analogy to TDTR) to investigate the cross-plane thermal conductivity of layered systems. Two extreme structures serve as showcases, on the one hand isotopically modulated silicon multilayers [25,30] and on the other hand Mo/Si ML, respectively silicide(SiMo)/Si multilayer (ML) [31,32]. While the first have an overall high conductivity and low interface defect density, the latter have a high acoustic impedance mismatch and (due to sputter growth) more defects. It is demonstrated that the 'effective' thermal conductivity of a layer stack can be determined in a wide range of values and separated from the unavoidable thermal interface resistance (TIR) at the interface between the sensing transducer layer and the investigated layer stack.

2. Materials and Methods

Sample preparation: The silicon-based multilayers on the basis of isotopically pure precursors (atomic mass Z_{Si} = 28, 29, 30 a. u.) were epitaxially grown by means of molecular beam epitaxy (MBE), respectively chemical vapor deposition (CVD, for the epilayer) on (100) oriented Si wafers of natural isotopic abundance (p-type substrates, boron doped, specific resistivity of 0.02 and 4 Ωcm, for the Z_{Si} = 29 and Z_{Si} = 30, respectively). The thickness of each individual isotope layer was set to 10 nm resulting in a total thickness of about 400 nm for the entire layer structure with 20 repeat structures. A reference structure of 400 nm silicon of natural abundance was grown as reference sample. In all cases the MBE, resp. CVD process started with the growth of an about 100 nm thick natural silicon buffer layer before the isotopic layer structure was deposited. Isotopic layering was verified by secondary ion mass spectroscopy (SIMS) [30], which gave a interface sharpness of <1 nm, defined by the precision in SIMS. The layers show a full epitaxy. X-ray truncation rod scattering at the (400) reflection (instrument SCD at KARA, KIT Karlsruhe, Germany) showed a modulation of intensity at the scattering vector shift corresponding to the periodicity, which implied a (9.7 + 9.7) nm layering with an isotopic strain of $\pm 1.9 \times 10^{-5}$, comparable to silicon isotopic lattice parameter differences [29]. The layer structure was verified by X-ray reflectivity at the synchrotron beamline SCD at KARA at an X-ray energy of 8.91 keV on a 6-circle diffractometer, using a linear pixel detector (Mythen, Dectris Ltd., Baden-Daettwil, Switzerland) as detection system.

Mo/Si and $Mo_{0.85}Si_{0.15}$/Si multilayers with 20 periods were deposited by magnetron sputter deposition. The experimental details for deposition of a single bilayer were reported in [32]. Mo/Si bilayer systems consist of an amorphous Si (a-Si) layer, and an amorphous or crystalline Mo layer. Under the present growth conditions, it was found that Mo crystallizes at a nominal thickness d(Mo) = 2.8 nm. $Mo_{0.85}Si_{0.15}$/Si shows a similar amorphous-crystalline transition at d(MoSi) = 4.0 nm. Due to the formation of an amorphous silicide interlayer with d $\simeq$ 0.5–1 nm, in both cases the crystalline thickness is slightly lower than the nominal deposited layer thickness. For Mo/Si multilayers, silicide interlayers were found at both interfaces. The studied multilayers with sublayer ratio of 0.5 ± 0.05 were deposited on Si(100) wafers covered by native oxide, and capped with a Si layer (typical thickness 3 nm). The period of the multilayers was varied. The Mo deposition rate was 0.0307 nm/s, the Si deposition rate was 0.0125 nm/s. For the Mo/Si sample with 18.2 nm period, the Si rate was increased to 0.021 nm/s. The $Mo_{0.85}Si_{0.15}$/Si alloy layers were co-deposited from both targets. The Si deposition rate was reduced to 0.0042 nm/s, resulting in a total deposition rate of 0.0349 nm/s for $Mo_{0.85}Si_{0.15}$.

Figure 1 shows XRR measurements of MoSi/Si (black dots) and $Mo_{0.85}Si_{0.15}$/Si (blue circles) multilayers with different period. For all samples, the higher order ML peaks are significantly broadened and dampened. This is mainly due to the ripple structure of the layers, which increases from period to period and is strongly correlated [33]. Comparing the Mo/Si and $Mo_{0.85}Si_{0.15}$/Si multilayers with 5.5 nm and 5.8 nm period, it is obvious that the multilayer peaks of Mo/Si are much more dampened. For bilayers, it was shown that below the crystallization threshold the roughness is comparable to the substrate roughness, but increases with layer thickness above the threshold. For all samples except of the 5.8 nm $Mo_{0.85}Si_{0.15}$/Si multilayer the Mo or $Mo_{0.85}Si_{0.15}$ layer thickness is above the crystallization threshold. The critical thickness d_c for crystallization is 2.8 nm for Mo and 4 nm for $Mo_{0.85}Si_{0.15}$ [32]. For the layers with period around 5.5 nm, the Mo layer is above d_c, while the MoSi layer is below d_c, explaining the larger damping of Mo. For 10 nm period, the difference is less pronounced since both layers are crystalline. The ML period and the Mo or MoSi fraction of the ML period were confirmed by fits of the measured XRR data. The fits were performed with the software GenX, using the Parratt formalism [34]. To simplify the stack model, it was assumed that the roughness increases linearly with the ML period. Each Mo or MoSi layer results in a stepwise roughness increase, but the silicon layers and the silicide interlayers only replicate the roughness of the layer below. Only for the Mo/Si ML with 5.5 nm period, the amorphous silicide interlayers at both interfaces were included in the fit. For all other samples, the interlayer signal was smeared out due to the layer corrugation and lower electron density contrast between $Mo_{0.85}Si_{0.15}$ and the silicide interlayer. For all samples, a surface roughness of the order of 1 nm was determined. The interlayer roughness increased up to 1–3 nm. These high values are caused by the ripple structure of the layers. However, a reliable quantification, based on the Parratt algorithm and a Gaussian roughness model is not possible any more since the roughness is correlated and comparable to the layer thickness.

Time-domain X-ray thermal scattering: one general approach of time-domain determination of thermal conductivity is to impulsively heat a transducer layer on top of the investigated layered structure and record its temporal cooling via an appropriate sensing probe. In established TDTR [20] a femtosecond laser is split between pump and probe pulses that can be delayed with respect to each other to probe the temporal temperature decay. An aluminum transducer usually serves as heater and temperature probe. Its optical reflectivity can be calibrated to derive the temperature. Aluminum shows a sufficiently linear thermal response for calibration. Laser focusing conditions (broad beams versus tight spots, or even thermal gratings) define whether the cross-plane conduction or also an in-plane conductivity component can be probed.

Pulsed X-ray scattering (and electron scattering) has shown to be a useful approach to determine thermal conductivity in layered structures [24,26,27,35,36]. In principle, TDXTS applies the same principles as TDTR above to determine the temperature of the top transducer via the thermal expansion coefficient of the material. While using synchrotron radiation for pulsed X-rays figures a rather involved approach, some advantages may concern the fact that thermal expansion can readily be quantified, elastic behavior can be distinguished from plastic deformation and the delay span of two independent sources can be extended from the picosecond to the millisecond range (as would also be possible with two independent lasers). Here we document the general approach and discuss the various aspects of real structures.

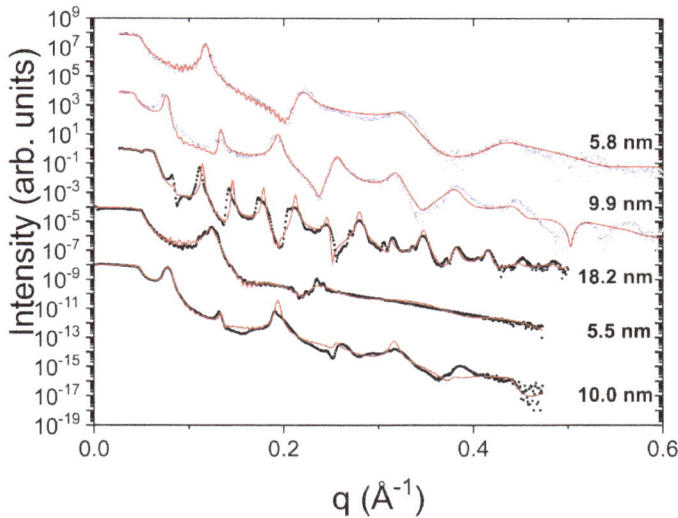

Figure 1. X-ray reflectivity measurements for the MoSi/Si (black dots) and $Mo_{0.85}Si_{0.15}$/Si (blue circles) multilayers. The period is indicated. The corresponding simulated XRR curves are shown as red lines.

The pump-probe experiment was realized at the beamline ID09 at the European Synchrotron Radiation Facility (ESRF, Grenoble, France). A single-line undulator (U17) produces an X-ray beam that is pulsed with the filling mode of the electron bunches in the ring at a center energy of 15.2 keV. The X-rays are focused by a toroidal mirror onto the sample position to form a $\simeq$0.1 mm sized focus, which elongates to 0.6 mm along the beam due to the shallow incidence. The frequency of the X-ray pulses arriving at the sample is reduced by a set of mechanical choppers (coarse heat load chopper and a microsecond high speed chopper at 986 Hz [37]). Further monochromatization is achieved by a Ge channel-cut crystal or a pair of multilayers (see discussion below). A femtosecond regenerative amplifier (Ti:Sa, Coherent Inc., Santa Clara, CA, USA) produces a laser pulse train synchronized to the X-ray pattern with a delay that can be modified in 5 ps steps up to 1 ms. The X-ray scattering is recorded on a (not time-resolving) CCD detector (MX170-HS, Rayonix LLC, Evanston, IL, USA) that accumulates the signal from typically 200 subsequent pulses at fixed delay after laser excitation. The acquisition is then repeated at a shifted delay to map the temporal evolution between delay $\tau = 0$ and <100 µs. The sketch in Figure 2 describes the approach. At a given laser fluence the position of laser-X-ray overlap is kept fixed on the sample surface.

Various metals can be employed as thermal transducer layers, as the thermal expansion from any suitable Bragg reflection of the layer can be recorded for determining the temperature. This can be done for single-crystalline transducers [24,38,39]. However, polycrystalline layers [25,36] simplify the alignment, as the incident Bragg angle has to be aligned only coarsely, while no scanning is required [40]. A typical image of the powder ring of gold is shown in Figure 2, as well as a radial profile as function of full scattering angle 2Θ (right side). We have investigated several materials to be used as transducers, such as silver, gold, platinum or aluminum. Gold has turned out to be the most suitable metal, as it has a high scattering cross section, is insensitive to atmospheric degradation and shows a strong crystalline texture. The latter represents the most important aspect as sputtered or evaporated gold thin films show preferential orientation of the lattice planes in <111> direction [41]. Evaporated films have displayed lateral peak widths of the (111) reflection of down to 2 degrees, while the sputtered films are typically more spread with a 8–10 degree orientation towards the surface normal. This orientation ensures a high scattering cross section with a selectivity towards cross-plane expansion. Silver is similarly suited, while displaying oxidation that may modify the properties and complicate modeling. The film thickness has to be balanced between thinnest films for achieving the highest time resolution and thus highest sensitivity on changes in the subsurface regions, in particular with overall high conductivity as in silicon, and a higher scattering cross section for thicker films. Additionally, thin films tend to de-wet at too high laser fluence. They also fail to fully localize the initial energy deposition in the metal transducer due to finite absorption cross section and injection of fast electrons into the sub-surface region [42,43]. We found an optimal gold thickness between 30 and 40 nm thickness. The layers used here were typically 44 nm thick.

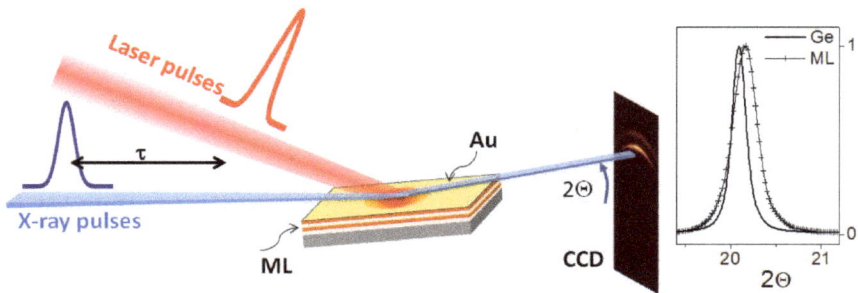

Figure 2. (**Left**) scheme of the laser pump, X-ray probe setup for the determination of cooling kinetics with 100 ps time resolution at a 3rd generation synchrotron. The train of X-ray and laser pulses are overlapped on the wafer surface with a defined delay τ between the pulses. Powder scattering is recorded on a CCD detector. Its angular deviation $2\Theta(\tau)$ serves as temperature monitor of the gold cover layer. (**Right**) Comparison of radial powder lines of the (111) reflection of gold in monochromatic X-ray mode (Ge monochromator) and in broad band mode (multilayer monochromator).

Usually monochromatic X-rays would be employed in order to achieve the highest resolution in Bragg angle change due to thermal expansion, which just amounts for a few tenths of a degree at 2Θ of $20°$. Nevertheless, the beamline ID09 offers using either a crystal monochromator (Ge channel cut in the present case) or a multilayer pair (consisting of Ru/B_4C). The energy resolution of the monochromator ML together with the sharp undulator edge at 15.2 keV [44] allow for the preparation of an X-ray beam of 1.3% bandwidth [45]. While this results in a broadening of the typically >0.1° reflections, the angular shift can still be resolved with the same or better precision due to higher counting statistics. Thus, a total of 2000 shots, or 2 s exposure time is sufficient for a temperature resolution below 0.5 K (see below). Data from the multilayers below is derived in both modes.

The near-infrared laser beam (800 nm, 2 ps) is focused to a size (0.5 mm) much larger than the X-ray footprint on the sample in order to achieve a homogeneous heating. Fluence is controlled by a motorized waveplate/polarizer combination to control absolute temperature rise. As discussed below, we restricted the fluence to a temperature rise of the gold film of below 150 K, typically only 80 K.

Calculation of heat transfer: The temporal cooling of the gold layer is related to energy loss due to heat conduction in contact to the underlying layered structure. Therefore radiation losses can be neglected. As such, the thermal conductivity of the structure can be deduced from the cooling speed. Ideally the analysis of the temporal behavior also allows for resolving depth-dependent properties. However, a proper modeling is required for a quantitative derivation. Here we assume a purely cross-plane heat flow, such that the systems are effectively modeled by 1-dimensional heat transfer. Importantly, any model has to incorporate the thermal interface resistance [19], as in particular the gold-semiconductor interfaces represent a strong obstacle against heat transfer. The so-called Kapitza resistance has been shown to be indispensable also for room temperature phenomena, as long as nanoscale processes are considered [23,46]. One approach is to solve the differential equations for diffusive heat transport by incorporating the effective heat flux at the boundaries of $\dot{Q} \propto A \cdot \Delta T$ with the surface area A and the temperature drop ΔT across the interface [30]. Here, we follow an approach using Laplace space, that allows for a analytic formulation of conduction in layered systems for specific starting conditions, such as a uniformly heated transducer layer [47]. The so-called transmission-line approach can be extended for multiple interfaces [29,48,49].

$$T(s) = \frac{Q(s)C_{11}}{e_1\sqrt{s}C_{21}},\tag{1}$$

with T being the temperature as function of the Laplace coordinate s, $Q(s)$ being the Laplace form of the temporal heating source term (the laser profile) and C being the matrix product of (2×2) matrices containing the layers' material parameters in the form $e_i = \sqrt{\rho_i c_i \kappa_i}$, the layer thickness d_i and the TIR between each layer (with ρ density, c specific heat and κ thermal conductivity).

Individual layers are modeled by thermal conductance and diffusivity in a (2×2) matrix formalism. TIR can be introduced in a similar way. With this parametrization, the temperature in Laplace space is calculated by multiplying the matrices and deriving the temporal temperature decay of the transducer layer by backtransform. We use 3 or 4 layers, (1) gold top layer, (2–3) multilayer stack, and (4) substrate (semi-infinite). For details see [48,49].

Both the output of the Laplace approach and the solution of the diffusion equations scale with the absolute change in temperature through the source term. Explicit temperature dependence only enters via temperature-dependent material parameters, such as diffusivity or thermal expansion coefficient. Thus, a direct temperature dependence is not included in the Laplace approach.

In general, ρ, c, κ, d and the TIR could be free parameters in a fitting procedure. This would, however, be highly underdetermined, as for instance, the TIR and k are strongly interrelated parameters. In practice, $\kappa_{Si,nat}$ is taken from literature, the thicknesses of the layers are determined independently through X-ray reflectivity, respectively crystal truncation rod scattering and the densities are taken as bulk values. The TIR between gold and silicon for the isotope ML is determined from the reference sample. Therefore only one free parameter, the change of thermal conductivity of the ML stack relative to natural silicon remains to be optimized. The mean square error between the differences between reference sample and multilayer sample and the simulations of a native substrate and the effective conductivity of the simulated multilayer is taken as figure of merit. For the MoSi layers the TIR and the effective κ are fitted simultaneously, which is possible due to the strongly different time scales of cooling ruled by the TIR and the conductivity of the layer stack. As consequence, the determination of absolute values of the conductivity is less precise than the relative change between a reference sample and the investigated layer structure.

Static characterization of thermal expansion coefficients: The thermal expansion coefficient has been tested on a bulk sample, nanoparticles and the investigated films. The gold plate (99.95%, Chempur, Karlsruhe, Germany) as well as the gold nanoparticles were measured on a lab diffractometer with Cu anode(Rigaku Corp., Tokyo, Japan) equipped with a wafer heater (DHS1100, Anton Paar, Graz, Austria) and a linear detector (D/teX Ultra, Rigaku Corp.). The nanoparticles consisted of commercial spheres of 100 nm (BBInternational, Crumlin, UK) deposited in a single layer on a silicon wafer [50]. The organic coverage between substrate and particles prevents a lattice strain between particles and substrate. The gold film (44 nm) on top of the reference silicon wafer was analyzed at the beamline SCD, KARA by using a custom-made resistive heater with temperature control (331, Lake Shore Cryotronics, Inc., Westerville, OH, USA) and a linear detector (Mythen). In both cases, the temperature was ramped linearly at 1 K/min while recording the powder profile of the (111) reflection. The change in scattering angle is then converted to relative lattice expansion.

3. Results and Discussion

3.1. Thermal Expansion of Strained Thin Films

While the simulation of the thermal conductivity is inherently not temperature-dependent, the absolute temperature determination is nevertheless important for estimation of non-linear contributions of temperature on thermal strain. This could be a non-linear expansion coefficient, as well as plastic deformation. We have determined the lattice expansion coefficient of a bulk gold plate, gold nanoparticles and the used gold transducer layers. The thermal expansion coefficient of bulk gold is well known, while different parametrization yields different values. We have compared the lattice expansion of the gold plate in Figure 3 to two expressions given by Touloukian [51] and in ref. [52]. While the deviation from linearity is low, the power law parametrization to second order by Suh et al. fits our results better. In general, the expected expansion is well reproduced. We expect a similar behavior for finite-size gold particles, while a change of absolute expansion may take place for small particles due to modified interatomic potentials. This effect, however, is small if the particles size exceeds some tens of nanometers in gold [40]. The nanoparticles of 100 nm size and high crystallinity investigated here [53] show a thermal expansion coefficient that is well comparable to that of bulk gold (Figure 3, middle). This is reasonable, as due to the absorption process on the silicon surface the particles are coupled to the substrate only via organic ligand molecules, thus no epitaxial strain is supposed to occur. The cooling cycle in both cases is fully reversible.

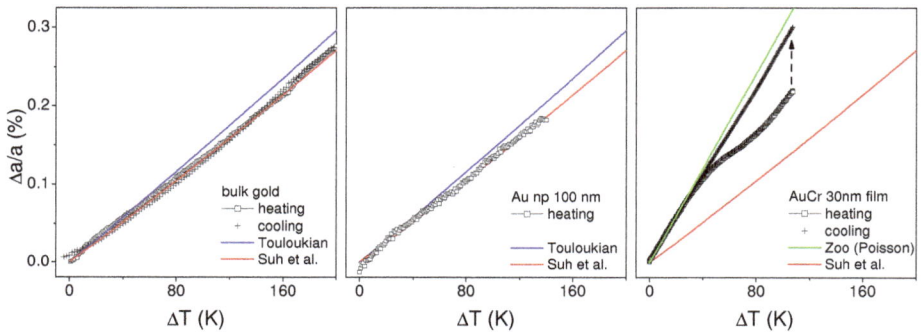

Figure 3. (**Left**) relative thermal expansion of a bulk gold plate as function of temperature rise above room temperature. The symbols mark heating ramp (dots) and cooling ramp (crosses). The lines are the calculated thermal expansion according to refs. [52] and [51]; (**Center**) relative thermal expansion of a single layer of spherical gold nanoparticles on a silicon wafer as function of temperature rise together with the calculated thermal expansion as in the left figure; (**Right**) relative thermal expansion of a sputtered gold film on silicon. The cooling curve is shifted for clarity to force overlap with the heating curve at $\Delta T = 0$ K (indicated by the arrow). The lines are the calculated bulk thermal expansion from [52] as well as a calculation assuming additional Poisson expansion [54].

The situation is different for a thin metal layer on a flat substrate. Although the growth via evaporation or sputtering does not result in an epitaxial relationship between the (oxidized) silicon substrate and the gold lattice, a strain is set to happen simply because the in-plane expansion needs to meet that of silicon, if no delamination is occurring. In fact, we used a thin chromium layer between silicon and gold for enhancing adhesion and lowering the TIR [41,55]. The out-of-plane expansion will be significantly modified, if the substrate possesses a significantly different expansion coefficient and the layer is clamped to it. The modified expansion coefficient can be rationalized as a compensation of in-plane restriction by Poisson deformation in the out-of-plane direction. Using the Poisson ratio ν the out-of-plane expansion α_{film} on a homogeneously heated layer system can be described as [54]:

$$\alpha_{film} = \frac{\alpha_{gold} - \alpha_{Si}}{1 - \nu} \qquad (2)$$

Given a room-temperature thermal expansion of silicon (3×10^{-6}/K), which is a factor of 5 smaller than that of gold (13.7×10^{-6}/K in the present case), it is expected that the gold film will expand in excess in the out-of-plane direction. Indeed, we observe an initial coefficient of 3×10^{-5}/K for the gold film, almost as high as the prediction in Equation (2). Earlier, a slight relaxation in in-plane direction was been observed [41], that may have been caused by the granular structure of the film.

The initial linear relation of lattice expansion with temperature is found to deviate towards a smaller slope at temperatures as low as 50 °C. At higher temperature the initial slope is reached again. The subsequent cooling ramp, on the other hand, shows a good linear relationship. Such effects are seen prominently during the first temperature cycle, while they largely diminish in subsequent cycles (not shown). Therefore we attribute this behavior to plastic deformation, respectively annealing of defects during growth. It is a common observation that polycrystalline films are under in-plane compressive strain. This strain can relax upon annealing to reduce the out-of-plane expansion [56]. Practically, such a non-linear relation would be a strong hindrance for using such layers as temperature sensors (including possibly TDTR). In following we will show that the effect is under control, when (i) limiting the temperature rise and (ii) correcting for the plastic deformation, which gradually builds up with laser exposure. The plastic deformation is quantified during the time-resolved measurements with interleaved measurements at negative delay.

For time-resolved heating, a similar uni-axial expansion of the gold film is expected, with an even higher out-of-plane coefficient, given that the silicon substrate heats up even less due to fast heat dissipation.

For photo-excitation conditions of a gold film Nicoul et al. [36] have given a relationship between temperature rise based on the model by Thomsen [57], which suggests an even higher increase of the out-of-plane expansion coefficient, which we, however, can not confirm.

3.2. Heat Transfer in Epitaxial Isotopically Modulated Multilayers

A typical decay of the thermal lattice expansion following photo-excitation is displayed in Figure 4. For negative delays, the probe pulse precedes the laser pump pulse so that no expansion is observed. This changes while the delay enters the pulse length of the X-ray pulse, where a gradual increase of expansion is seen that saturates at a 100 ps delay. Thereafter, the expansion decays with a particular characteristics. The first 1–2 ns delay span is described by an exponential decay with a lifetime around 1.2 ns (see Figure 5), followed by a slowing down of relaxation (and thus cooling). The final decay after some 100 nanoseconds approaches a square-root form. The exponential decay is explained by the predominance of the TIR in a situation, where the gold film is warm, while the silicon substrate is still cold. In the TIR case the absolute flux is proportional to the difference in temperature, thus an integration yields an exponential decay. The square-root behavior is clearly linked to bulk diffusion, as found in Fick's law. On the nanosecond time scale the two cooling regimes are balanced.

At the same time, a measurement of expansion at negative decay before each positive delay point shows a variation in absolute value, which would not be expected for a fully reversible situation. The delay axis for these reference measurements is in fact an exposure scale (as indicated on the top axis). With ongoing exposure the expansion first turns negative, while at a later stage is becomes positive again. Keep in mind that the each data point encompasses 5000 individual laser pulses, but at fixed fluence, so that the end temperature is reached thousands of times with ongoing exposure. In other samples the detailed course of the irreversible changes would differ in its exact shape. Therefore we interpret this exposure behavior as a plastic deformation in a similar fashion as observed in the steady-state heating above. This is rationalized by the notion that expansion at positive delay follows that at negative delay for the last measurement points, where the reversible heating of the gold film has decayed. In both cases the plastic deformation remains. An effect by a slow warming of the sample by the series of laser pulses is also possible, but accounts only for several Kelvin at 1 kHz repetition rate [58]. Additionally, the equilibrium will be reached within the first hundreds of laser shots, so that a drift in the expansion baseline (−2 ns) would occur within the first few data points. In order to derive the true time-resolved expansion the contribution due to plastic deformation is subtracted.

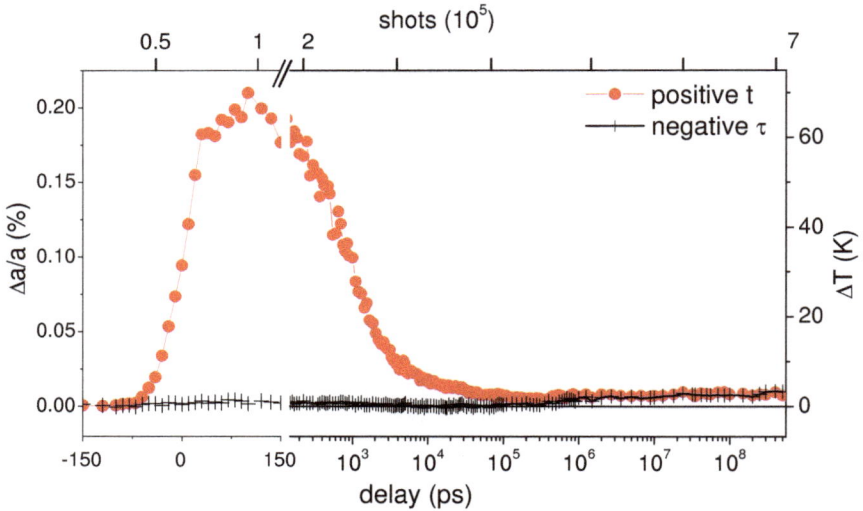

Figure 4. Lattice thermal expansion of a gold film on top of a natural silicon wafer as function of delay between laser and X-ray probe pulse. The circles mark an acquisition, where the delay has been constantly increased after each measuring point, while the delay has been kept constant at −2 ns for the data with crosses, but directly preceding the acquisition with varied delay. The corresponding number of laser shots accumulated on the illuminated spot is marked on the top axis.

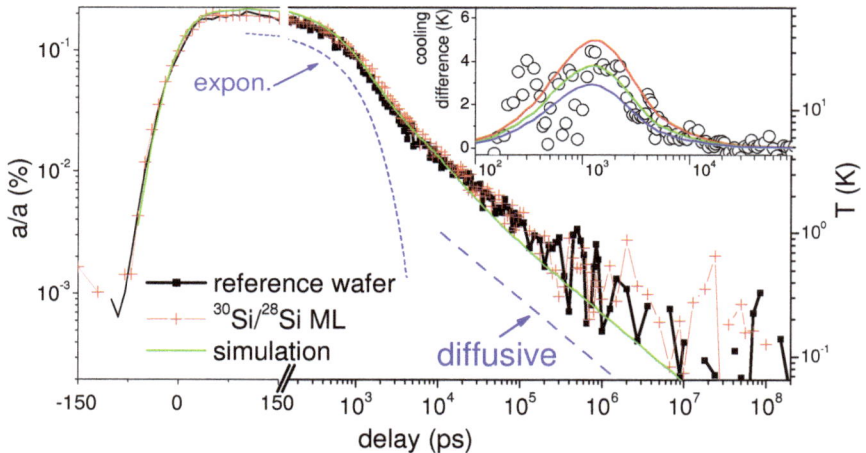

Figure 5. Lattice expansion of the gold layer on top of two silicon samples, the reference substrate (black squares) and the wafer with the isotope-modulated multilayer stack (red crosses). The lines are a guide to exponential and diffusive decay, respectively simulations of temperature decay based on the layer model in conductivity. The shape around $\tau = 0$ is modeled by a sigmoidal function with 30 ps Gaussian width. The inset displays the change in cooling of the ^{30}Si/^{28}Si-ML compared to the native Si reference sample. The lines are simulations by assuming a change of thermal conductivity to 55 (red), 60 (green) and 65 (blue) W/(m·K), from top.

This correction yields a precise decay of lattice heating shown in Figure 5 where the late-time square-root decay can be reproduced in good agreement to the calculations. The axis on the right indicates the absolute temperature change according to Equation (2). Figure 5 displays both the cooling of the native silicon reference sample as well as that of a $20 \times (10 + 10)$ nm isotope-modulated stack of ^{28}Si and ^{30}Si. In both cases the initial exponential as well as the final square-root decay coincide. This is rationalized by a comparable TIR value of both samples as well as similar bulk heat diffusion into the (same) substrate. As consequence, it is possible to fix the TIR value in both cases to extract the change in effective thermal conductivity of the stack as single parameter. Cooling is delayed slightly for the multilayered sample, which qualitatively points towards a lower conductivity. Nevertheless, a quantitative determination relies on the modeling of heat transfer. The best value for the multilayer conductivity is (61 ± 10) W/(m·K) as compared to that of the reference sample of (130 ± 25) W/(m·K). Earlier, we have reported slightly different values in the present system (compare also Table 1). We attribute this to the different gold film thickness. Here, we have chosen 44 nm thick layers as compared to the <30 nm [43] or 80 nm [25] earlier. This avoids on the one hand leaking of fast electrons (or laser light) into the silicon stack and on the other hand sensitivity to absolute high values of thermal conductivity is preserved by keeping the gold layer as thin as possible.

An in-depth discussion of the interpretation of thermal conductivity in complex tailored systems is beyond the scope of the present work. In general, heat can be transported by the conduction electrons as well as phonons. In our low-doped silicon samples electron conduction at room temperature plays a minor role. Limitations to heat conduction by phonons are given by several effects that limit the mean-free path of phonons and thus represent an obstacle to transport. The conventional mechanisms described are incoherent ones, where a phonon is scattered at point defects or an interface or by multiphonon process (Umklapp processes). In well-ordered multilayers, on the other hand coherent phonon interaction across the interfaces can also take place. These are described to cause partial phonon band blockage or even Bloch-like pass bands. Recent lattice dynamics also point towards mini-Umklapp processes in periodic multilayers [29].

Simulations of the cross-over between coherent and incoherent transport in the present silicon isotope multilayers shall illustrate the expected behavior. Frieling et al. have conducted non-equilibrium molecular dynamics simulations (NE-MD) of heat transfer across a stack of mass-modulated silicon layers [59,60]. NE-MD calculations of the thermal conductivity of natural Si were also performed for comparison. Details on the calculations and on the MD simulation cell are given in Ref. [59].

Figure 6 reproduces the main results. The effective thermal conductivity starts by dropping from the bulk value of 112 W/(m·K) at very large period, thus only a few interfaces are sensed by the phonons. The simulated bulk conductivity is lower than the experimental one due to the finite size of the simulation box [60]. When reducing the period, the conductivity drops, essentially because the interface density increases. The product of TIR at each interface and the layer thickness qualitatively reproduces this monotonously decaying function with reduced period. The TIR value has been taken from [30]. However, below a period of 6 nm the conductivity rises again for the ideal ML with perfect interfaces. On this length scale the phonon band structure for long-wavelength phonons does not sense the periodicity any more, thus marking a transition from incoherent to coherent transport.

On the other hand, point defects scatter phonons effectively as well. This can already be seen for the structure with gradual mixing between the layers (for details see [59]). When rearranging all the isotopes in the layers into arbitrary positions (which has been done experimentally by annealing the ML at 950 °C for 120 h) one obtains a homogeneous alloy, however with disordered isotope distribution. In this situation the simulated thermal conductivity drops further.

Comparing the prediction to our data we find that the drop in conductivity from the natural silicon is higher than in the MD, but follows the same order of magnitude, the drop for the ^{30}Si/^{28}Si being higher than that of the ^{29}Si/^{28}Si ML. Forming an alloy from these layers additionally reduces conductivity further (see also Table 1). Interestingly this further drop is less pronounced in the experiment than predicted by the MD simulations. This points towards defects already playing a role for the MBE-grown layers as compared to the ideal situation in MD.

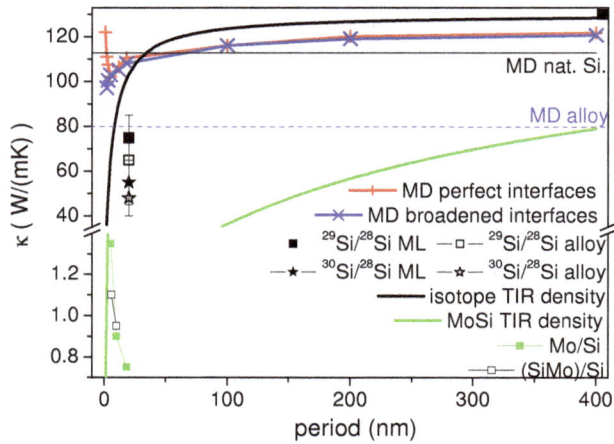

Figure 6. Comparison of the measured thermal conductivities of the silicon isotope multilayers of molecular dynamics calculations as a function of the periodicity of the stack (Frieling et al., [59]). The horizontal lines mark the simulation result for natural silicon (black line) and for the random $^{28}Si_{0.5}/^{30}Si_{0.5}$ alloy (blue dashed line). The lower part of the figure contains experimental data from the Si/Mo multilayers. The curved lines indicate an effective conductivity as a sum of TIR of the individual layers for the isotope modulation and the Si/Mo stacks (lower green thick lines), respectively.

Table 1. List of all used thin-film structures with the multilayer period, TIR at the interface gold-layer and the determined effective thermal conductivity of the full layer stack.

Sample	Period	TIR (m·K/W)	κ_1 (W/(m·K))
nat-Si	–	4.5×10^{-9}	130
28-Si/29-Si	20 nm	4.5×10^{-9}	(81 ± 10)
28-Si/29-Si	alloy	4.5×10^{-9}	(79 ± 10)
28-Si/30-Si	19.4 nm	4.5×10^{-9}	(61 ± 10)
28-Si/30-Si	alloy	4.5×10^{-9}	(51 ± 8)
100 nm Mo	–	2.5×10^{-8}	(80 ± 20)
Mo/Si	5.5 nm	2×10^{-8}	(1.35 ± 0.2)
Mo/Si	10 nm	2×10^{-8}	(0.9 ± 0.15)
Mo/Si	18.2 nm	2.2×10^{-8}	(0.75 ± 0.15)
(SiMo)/Si	5.8 nm	2×10^{-8}	(1.1 ± 0.2)
(SiMo)/Si	9.9 nm	2×10^{-8}	(0.95 ± 0.2)

The precision of the evaluation of conductivity is naturally limited by the resolution in temperature rise and thus by counting statistics of the scattering yield (see inset to Figure 5). Additionally, the high absolute conductivity of the system poses a particular problem to the time-domain approach. As the quantification of conductivity is based on determining the cooling rate, it is limited by other sources of thermal resistance in the system. The first and most important one is the TIR at the interface gold-layer, which limits heat transfer. With a high TIR, the changes in cooling at later times than the exponential decay in Figure 5 become marginal. A similar influence may be imposed by the conductivity of the substrate. On a substrate with very low conductivity the heat may be kept in the layer system, modifying the residence time. In the current approach the cooling was followed over several decades in delay, such that resistance may be localized in depth to a certain extent. Nevertheless, determination of conductivities above 100–150 W/(m·K) is restricted with the present setup, unless a lower TIR at the gold-layer interface can be achieved. At the same time, lower TIR often correlates with lower Schottky barrier, such that non-thermal electrons might escape into the layer system. In that case, the thermal conduction would have to be modeled in a more detailed way.

A critical assessment of absolute and relative errors of the evaluation of κ reveals that relative changes of analogous samples as listed in Table 1 can be rather small, while the absolute determination of the conductivity of the reference sample with natural silicon is strongly related to other optimization parameters, such as the TIR. Earlier, a direct solution of the differential equations of thermal transport has been used [43], which allowed to incorporate direct energy deposition into the layer by a leaky gold film or fast electrons. There, the TIR was found to be lower, while the conductivity of native silicon was higher. Still, the relative reduction of conductivity in the multilayers was similar.

3.3. Heat Transfer in Sputtered Molybdenum Silicon Multilayers

Mo/Si multilayers represent the opposite case as compared to the isotope layers, which show low density contrast across the interfaces. The density difference between molybdenum and silicon results in a much lower effective thermal conductivity of the layer system. Additionally, interlayers of amorphous silicon, respectively silicide may reduce thermal conductivity further. Bozorg-Grayeli et al. [31] investigated Si-Mo multilayer structures to find a thermal conductivity as low as 1.1 W/(m·K) and additionally observed the increase of conductivity of a molybdenum silicide film with annealing from 2 to 3 W/(m·K) as following the irreversible amorphous-crystalline transition. Furthermore the conductivity of such multilayers has been investigated in terms of cross-coupling between electronic and phononic sub-systems [61], which opens a further pathways for heat conduction. A detailed model predicted 1.3–1.5 W/(m·K), which matches the observed 1.2–1.4 W/(m·K).

The TDXTS data (Figure 7) on the set of multilayer structures investigated here shows that cooling in general is much slower for the MoSi system than for the isotope samples with the limit of bulk conductivity (as seen by the change from the power-law to the diffusive limit) not being reached before 1 μs. The thermal decay of the multilayers is found to be an order of magnitude slower than the decay of the molybdenum layer. Results of the calculations are found in Table 1. The 100 nm molybdenum layer shows a conductivity 80 W/(m·K), again limited in precision by the large TIR, comparable to tabulated values. In contrast, the conductivity of the ML is lowered to 0.75 to 1.35 W/(m·K). A fit with a single ML layer with effective conductivity is only reproducing the time scale, while some deviations are still seen Figure 7. A better fit can be achieved by dividing the ML thickness in two equal regions and allowing the conductivity to vary independently. In that case the model fit is much better, but in all cases suggests that the lower part of the ML has a higher conductivity (of 2–4 W/(m·K)) than the top part. This points towards the increase in disorder and ripple effect disturbing the layering and reducing effective conductivity.

Overall, it is found that the thermal conductivity is lowest for the larger periods as well as for the silicide ML. A generic diffuse mismatch mode with counting interfaces would predict the opposite. The comparison with an effective conductivity for a typical TIR of (heavy) metal-silicon interfaces [41,62] of 10^{-9} m^2·K/W in Figure 6 shows that the effective conductivity is in the same range, while the thickness dependence is not reproduced. However, a periodicity of 6, respectively 10 nm, can already be too short for having an effect on long-wavelength phonons, as has been predicted by theoretical considerations [15,63] and simulations [59,64]. More likely, in view of the lower conductivity of the silicide ML, the real structure plays a central role in the magnitude of suppression of heat conduction. For the lowest periods the intermixing at the interface and the lateral variation in periodicity can lead to an effectively higher conductivity. The silicide ML in general show a better definition of density variation. Despite the lower nominal TIR at each (SiMo)-Si interface the effective conductivity is lower.

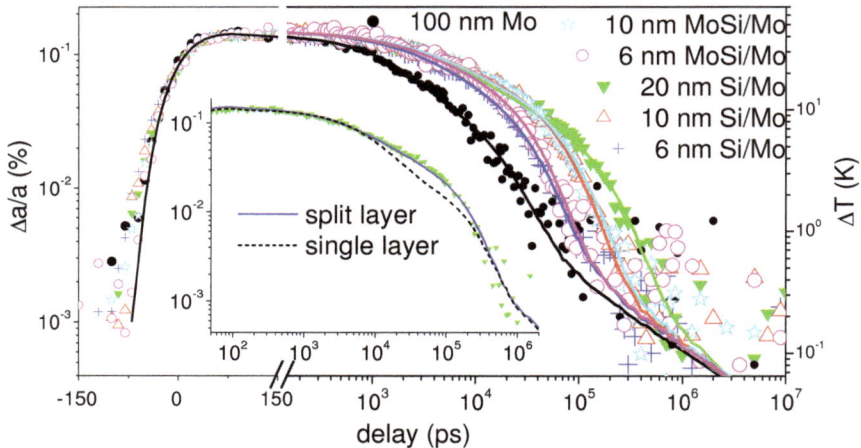

Figure 7. Lattice expansion of the gold layer on top of Si-Mo multilayer samples, as well as a 100 nm molybdenum layer on silicon substrate (black circles). The lines are results of a best fit with the heat transfer model with an effective thermal conductivity modeled independently for two splitted parts of the stack. The inset compares the two-layer fit with a single-layer fit.

4. Conclusions and Outlook

It has been demonstrated that time-resolved X-ray powder scattering of thermal transducer layers on top of a layered surface can be used to resolve and quantify the cross-plane thermal conductivity of the materials. It is possible to discern between TIR at the transducer interface and contributions from thermal conductivity. Synchrotron-based diffraction with sub-nanosecond resolution is particularly suited to follow thermal kinetics over several decades in time and temperature, thus allowing address as well the depth of where a resistance in heat flow occurs. While cooling is a diffusive phenomenon a limited depth resolution may still be achieved.

The methodology is analogous to the established TDTR, using a purely optical pump-probe approach. Meanwhile, in X-ray scattering understanding the atomic scale structure, plastic deformation processes and obtaining a absolute temperature calibration are straightforward. The delay range spans 100 ps up to millisecond times, which allows visualizing different heat conduction regimes. The transducer film has to be optimized in order to minimize the TIR at the interface to the probed thin films as well as in thickness. Thinnest films below 30 nm do not dissipate the total laser energy in the film, while thicker films lengthen the time scale for heat transfer into the thin films and thus reduce sensitivity for high conductivity in the probed structures. Gold has been found to be ideal due to its strong preferential crystal orientation for various growth methods, such as thermal evaporation or sputtering.

Nevertheless, the described method relies on access to highly specialized synchrotron beamlines. It would possibly not be used for routine characterization, but rather for selected advanced problems. A simpler approach may meanwhile be accessible on any synchrotron beamline, taking advantage of developments in detector technology [65] and data processing. With avalanche photo-detectors a time resolution on the nanosecond scale can be achieved [66,67]. This would allow recording shifts in powder peak position of the gold layer without a dedicated pump-probe beamline. In a pump-record approach the laser pump pulse represents the start signal for a time-resolved acquisition of scattering intensity while the x-ray emitter (also preferably a synchrotron) will serve as continuous source. A time-resolved linear detector will allow for a resolution of the powder profile, while a point detector with knife-edge discrimination may already suffice to quantify the amount of peak shift and thus lattice expansion. At the same time, the pump-record approach would minimize the influence of plastic deformation in a similar way as the lock-in approach of TDTR. Even laboratory sources based on plasma-generated X-rays or liquid anodes may deliver sufficient pulsed flux for meaningful data collection [10,68].

Author Contributions: The experimental design has been conceived by H.B., T.B. and A.P., the experiments on static heating have been performed by G.B., A.P. and M.Z., while time-resolved experiments have been conducted by H.B, S.E. and A.P. The data has been discussed among all authors. Samples were grown by B.K. (MoSi multilayer), respectively received from U. Aarhus. The data analysis has been done by B.K. (reflectivity), M.Z. (static heating), S.E., C.G., H.B. and A.P. (conductivity simulations). The article has been written with contributions from all authors.

Funding: This work was funded by the Deutsche Forschungsgemeinschaft in the frame of the priority program SPP1386 'Nanostructured Thermoelectric Materials: Theory, Model Systems and Controlled Synthesis' (BR 1520/13-1 and BR 1520/15-1) and individual grants for AP and within the Heisenberg program for HB (BR 1520/10-2). The isotopically enriched Si was developed by the Initiatives for Proliferation Prevention Program of the Office of Nonproliferation Research and Engineering (NN-20) of the U.S. Department of Energy under contract DE-AC03-76SF00098. Part of the research is supported by the topic "From Matter to Materials and Life" within the Helmholtz Association.

Acknowledgments: Provision of beamtime at the facilities ANKA, later KARA (KIT) and ESRF is gratefully acknowledged. We wish to thank for the excellent support by D. Khakulin, M. Wulff and M. N. Pedersen. We are grateful for receiving MBE-grown isotope multilayers from U. Aarhus (J. Lundsgaard Hansen, A. Nylandsted Larsen) and the silicon isotopes from the Lawrence Berkeley National Laboratory (J. W. Ager III, E. E. Haller). Contributions in the early phase of the work from D. Issenmann are acknowledged.

Conflicts of Interest: The authors declare no conflict of interest.

Abbreviations

The following abbreviations are used in this manuscript:

TDTR	Time-Domain Thermal Reflection
TDXTS	Time-domain X-ray Thermal Scattering
XRD	X-ray Diffraction
XRR	X-ray Reflectivity
ML	Multilayer
d_c	critical thickness for crystallization
TIR	thermal interface resistance
κ	(effective) thermal conductivity
α	linear thermal expansion coefficient
ν	Poisson ratio
CVD	chemical vapor deposition
MBE	molecular beam epitaxy
NE-MD	non-equilibrium molecular dynamics simulations

References

1. Hicks, L.D.; Dresselhaus, M.S. Effect of quantum-well structures on the thermoelectric figure of merit. *Phys. Rev. B* **1993**, *47*, 12727. [CrossRef]
2. Snyder, G.J.; Toberer, E.S. Complex thermoelectric materials. *Nat. Mater.* **2008**, *7*, 105–114. [CrossRef] [PubMed]
3. Venkatasubramanian, R. Lattice thermal conductivity reduction and phonon localizationlike behavior in superlattice structures. *Phys. Rev. B* **2000**, *61*, 3091. [CrossRef]
4. Luckyanova, M.N.; Garg, J.; Esfarjani, K.; Jandl, A.; Bulsara, M.T.; Aaron, J.; Schmidt, A.J.M.; Chen, S.; Dresselhaus, M.S.; Ren, Z.; et al. Coherent Phonon Heat Conduction in Superlattices. *Science* **2012**, *338*, 936–939. [CrossRef] [PubMed]
5. Mukherjee, S.; Givan, U.; Senz, S.; Bergeron, A.; Francoeur, S.; de la Mata, M.; Arbiol, J.; Sekiguchi, T.; Itoh, K.M.; Isheim, D.; et al. Phonon Engineering in Isotopically Disordered Silicon Nanowires. *Nano Lett.* **2015**, *15*, 3885–3893. [CrossRef] [PubMed]

6. Asen-Palmer, M.; Bartkowski, K.; Gmelin, E.; Cardona, M.; Zhernov, A.V.; Inyushkin, A.T.; Ozhogin, V.I.; Itoh, K.M.; Haller, E.E. Thermal conductivity of germanium crystals with different isotopic compositions. *Phys. Rev. B* **1997**, *56*, 9431. [CrossRef]

7. Cahill, D.G.; Watanabe, F. Thermal conductivity of isotopically pure and Ge-doped Si epitaxial layers from 300 to 550 K. *Phys. Rev. B* **2004**, *70*, 235322. [CrossRef]

8. Cheaito, R.; Duda, J.C.; Beechem, T.E.; Hattar, K.; Ihlefeld, J.F.; Medlin, D.L.; Rodriguez, M.A.; Piekos, M.J.; Hopkins, P.E. Experimental investigation of size effects on the thermal Conductivity of Silicon-Germanium alloy thin films. *Phys. Rev. Lett.* **2012**, *109*, 195901. [CrossRef]

9. Gross, P.; Ramakrishna, V.; Vilallonga, E.; Rabitz, H.; Littman, M.; Lyon, S.A.; Shayegan, M. Optimally designed potentials for control of electron-wave scattering in semiconductor nanodevices. *Phys. Rev. B* **1994**, *49*, 11100. [CrossRef]

10. Bargheer, M.; Zhavoronkov, N.; Gritsai, Y.; Woo, J.C.; Kim, D.S.; Woerner, M.; Elsaesser, T. Coherent Atomic Motions in a Nanostructure Studied by Femtosecond X-ray Diffraction. *Science* **2004**, *306*, 1771–1773. [CrossRef]

11. Sondhauss, P.; Larsson, J.; Harbst, M.; Naylor, G.A.; Plech, A.; Scheidt, K.; Synnergren, O.; Wulff, M.; Wark, J.S. Picosecond X-Ray Studies of Coherent Folded Acoustic Phonons in a Multiple Quantum Well. *Phys. Rev. Lett.* **2005**, *94*, 125509. [CrossRef] [PubMed]

12. Tamura, S.; Tanaka, Y.; Maris, H.J. Phonon group velocity and thermal conduction in superlattices. *Phys. Rev. B* **1999**, *60*, 2627. [CrossRef]

13. Ezzahri, Y.; Grauby, S.; Rampnoux, J.; Michel, H.; Pernot, G.; Claeys, W.; Dilhaire, S.; Rossignol, C.; Zeng, G.; Shakouri, A. Coherent phonons in Si/SiGe superlattices. *Phys. Rev. B* **2007**, *75*, 195309. [CrossRef]

14. Bastian, G.; Vogelsang, A.; Schiffmann, C. Isotopic Superlattices for Perfect Phonon Reflection. *J. Electron. Mater.* **2010**, *39*, 1769–1771. [CrossRef]

15. Garg, J.; Bonini, N.; Marzari, N. High Thermal Conductivity in Short-Period Superlattices. *Nano Lett.* **2011**, *11*, 5135–5141. [CrossRef] [PubMed]

16. Ravichandran, J.; Yadav, A.K.; Cheaito, R.; Rossen, P.B.; Soukiassian, A.; Suresha, S.J.; Duda, J.C.; Foley, B.M.; Lee, C.H.; Zhu, Y.; et al. Crossover from incoherent to coherent phonon scattering in epitaxial oxide superlattices. *Nat. Mater.* **2014**, *13*, 168–172. [CrossRef]

17. Spitzer, J.; Ruf, T.; Cardona, M.; Dondl, W.; Schorer, R.; Abstreiter, G.; Haller, E.E. Raman scattering by optical phonons in isotopic $^{70}(Ge)_n$ $^{74}(Ge)_n$ superlattices. *Phys. Rev. Lett.* **1994**, *72*, 1565–1568. [CrossRef]

18. Cahill, D.G.; Fischer, H.E.; Klitsner, T.; Swartz, E.T.; Pohl, R.O. Thermal conductivity of thin films: Measurements and understanding. *J. Vac. Sci. Technol. A* **1989**, *7*, 1259–1266. [CrossRef]

19. Cahill, D.G.; Ford, W.K.; Goodson, K.E.; Mahan, G.D.; Majumdar, A.; Maris, H.J.; Merlin, R.; Phillpot, S.R. Nanoscale thermal transport. *J. Appl. Phys.* **2003**, *93*, 793–818. [CrossRef]

20. Cahill, D.G. Analysis of heat flow in layered structures for time-domain thermoreflectance. *Rev. Sci. Instrum.* **2004**, *75*, 5119–5123. [CrossRef]

21. Hu, M.; Hartland, G.V. Heat Dissipation for Au Particles in Aqueous Solution: Relaxation Time versus Size. *J. Phys. Chem. B* **2002**, *106*, 7029–7033. [CrossRef]

22. Plech, A.; Wulff, M.; Kuerbitz, S.; Berg, K.J.; Berg, G.; Graener, H.; Grésillon, S.; Kaempfe, M.; Feldmann, J.; von Plessen, G. Time-resolved X-ray diffraction on laser excited metal nanoparticles. *Europhys. Lett.* **2003**, *61*, 762. [CrossRef]

23. Plech, A.; Kotaidis, V.; Grésillon, S.; Dahmen, C.; von Plessen, G. Laser-Induced heating and melting of gold nanoparticles studied by time-resolved x-ray scattering. *Phys. Rev. B* **2004**, *70*, 195423. [CrossRef]

24. Shayduk, R.; Navirian, H.; Leitenberger, W.; Goldshteyn, J.; Vrejoiu, I.; Weinelt, M.; Gaal, P.; Herzog, M.; von Korff Schmising, C.; Bargheer, M. Nanoscale heat transport studied by high-resolution time-resolved x-ray diffraction. *New J. Phys.* **2011**, *13*, 093032. [CrossRef]

25. Bracht, H.; Wehmeier, N.; Eon, S.; Plech, A.; Issenmann, D.; Lundsgaard Hansen, J.; Nylandsted Larsen, A.; Ager, J., III; Haller, E. Reduced thermal conductivity of isotopically modulated silicon multilayer structures. *Appl. Phys. Lett.* **2012**, *101*, 064103. [CrossRef]

26. Harb, M.; von Korff Schmising, C.; Enquist, H.; Jurgilaitis, A.; Maximov, I.; Shvets, P.V.; Obraztsov, A.N.; Khakhulin, D.; Wulff, M.; Larsson, J. The c-axis thermal conductivity of graphite film of nanometer thickness measured by time resolved X-ray diffraction. *Appl. Phys. Lett.* **2012**, *101*, 233108. [CrossRef]

27. Shayduk, R.; Vonk, V.; Arndt, B.; Franz, D.; Strempfer, J.; Francoual, S.; Keller, T.F.; Spitzbart, T.; Stierle, A. Nanosecond laser pulse heating of a platinum surface studied by pump-probe X-ray diffraction. *Appl. Phys. Lett.* **2016**, *109*, 043107. [CrossRef]

28. Bojahr, A.; Herzog, M.; Mitzscherling, S.; Maerten, L.; Schick, D.; Goldshteyn, J.; Leitenberger, W.; Shayduk, R.; Gaal, P.; Bargheer, M. Brillouin scattering of visible and hard X-ray photons from optically synthesized phonon wavepackets. *Opt. Express* **2013**, *21*, 21188–21197. [CrossRef]

29. Issenmann, D.; Eon, S.; Bracht, H.; Hettich, M.; Dekorsy, T.; Buth, G.; Steininger, R.; Baumbach, T.; Lundsgaard Hansen, J.; Nylandsted Larsen, A.; et al. Ultrafast study of phonon transport in isotopically controlled semiconductor nanostructures. *Phys. Status Solidi* **2016**, *213*, 3020–3028. [CrossRef]

30. Bracht, H.; Eon, S.; Frieling, R.; Plech, A.; Issenmann, D.; Wolf, D.; Lundsgaard Hansen, J.; Nylandsted Larsen, A.; Ager, J., III; Haller, E.E. Thermal conductivity of isotopically controlled silicon nanostructures. *New J. Phys.* **2014**, *16*, 015021. [CrossRef]

31. Bozorg-Grayeli, E.; Li, Z.; Asheghi, M.; Delgado, G.; Pokrovsky, A.; Panzer, M.; Wack, D.; Goodson, K.E. Thermal conduction properties of Mo/Si multilayers for extreme ultraviolet optics. *J. Appl. Phys.* **2012**, *112*, 083504. [CrossRef]

32. Krause, B.; Abadias, G.; Michel, A.; Wochner, P.; Ibrahimkutty, S.; Baumbach, T. Direct Observation of the Thickness-Induced Crystallization and Stress Build-Up during Sputter-Deposition of Nanoscale Silicide Films. *ACS Appl. Mater. Interfaces* **2016**, *8*, 34888–34895. [CrossRef] [PubMed]

33. Voorma, H.; Louis, E.; Koster, N.B.; Bijkerk, F. Temperature induced diffusion in Mo/Si multilayer mirrors. *J. Appl. Phys.* **1998**, *83*, 4700–4708. [CrossRef]

34. Bjorck, M.; Andersson, G. GenX: an extensible X-ray reflectivity refinement program utilizing differential evolution. *J. Appl. Cryst.* **2007**, *40*, 1174–1178. [CrossRef]

35. Krenzer, B.; Janzen, A.; Zhou, P.; von der Linde, D.; von Hoegen, M.H. Thermal boundary conductance in heterostructures studied by ultrafast electron diffraction. *New J. Phys.* **2006**, *8*, 190. [CrossRef]

36. Nicoul, M.; Shymanovich, U.; Tarasevitch, A.; von der Linde, D.; Sokolowski-Tinten, K. Picosecond acoustic response of a laser-heated gold-film studied with time-resolved x-ray diffraction. *Appl. Phys. Lett.* **2011**, *98*, 191902. [CrossRef]

37. Cammarata, M.; Eybert, L.; Ewald, F.; Reichenbach, W.; Wulff, M.; Anfinrud, P.; Schotte, F.; Plech, A.; Kong, Q.; Lorenc, M.; et al. Optimized shutter train operation for high brightness synchrotron pump-probe experiment. *Rev. Sci. Instrum.* **2009**, *80*, 15101. [CrossRef]

38. Herzog, M.; Schick, D.; Gaal, P.; Shayduk, R.; Korff Schmising, C.; Bargheer, M. Analysis of ultrafast X-ray diffraction data in a linear-chain model of the lattice dynamics. *Appl. Phys. A* **2012**, *106*, 489–499. [CrossRef]

39. Pudell, J.; Maznev, A.A.; Herzog, M.; Kronseder, M.; Back, C.H.; Malinowski, G.; von Reppert, A.; Bargheer, M. Layer specific observation of slow thermal equilibration in ultrathin metallic nanostructures by femtosecond X-ray diffraction. *Nat. Commun.* **2018**, *9*, 3335. [CrossRef]

40. Plech, A.; Grésillon, S.; von Plessen, G.; Scheidt, K.; Naylor, G. Structural kinetics of laser-excited metal nanoparticles supported on a surface. *Chem. Phys.* **2004**, *299*, 183–191. [CrossRef]

41. Issenmann, D.; Eon, S.; Wehmeier, N.; Bracht, H.; Buth, G.; Ibrahimkutty, S.; Plech, A. Determination of nanoscale heat conductivity by time-resolved X-ray scattering. *Thin Solid Films* **2013**, *541*, 28–31. [CrossRef]

42. Hohlfeld, J.; Wellershoff, S.S.; Gudde, J.; Conrad, U.; Jahnke, V.; Matthias, E. Electron and lattice dynamics following optical excitation of metals. *Chem. Phys.* **2000**, *251*, 237–258. [CrossRef]

43. Eon, S.; Bracht, H.; Plech, A.; Lundsgaard Hansen, J.; Nylandsted Larsen, A.; Ager, J.W., III; Haller, E.E. Pump and probe measurements of thermal conductivity of isotopically controlled silicon nanostructures. *Phys. Status Solidi* **2016**, *213*, 541. [CrossRef]

44. Plech, A.; Randler, R.; Geis, A.; Wulff, M. Diffuse scattering from liquid solutions with white beam undulator radiation for photoexcitation studies. *J. Synchrotron Radiat.* **2002**, *9*, 287–292. [CrossRef]

45. Reich, S.; Letzel, A.; Menzel, A.; Kretzschmar, N.; Gökce, B.; Barcikowski, S.; Plech, A. Early appearance of crystalline nanoparticles in pulsed laser ablation in liquids dynamics. *Nanoscale* **2019**, *11*. [CrossRef]

46. Wilson, O.M.; Hu, X.; Cahill, D.G.; Braun, P.V. Colloidal metal particles as probes of nanoscale thermal transport in fluids. *Phys. Rev. B* **2002**, *66*, 224301. [CrossRef]

47. Ezzahri, Y.; Grauby, S.; Dilhaire, S.; Rampnoux, J.M.; Claeys, W.; Zhang, Y.; Shakouri, A. Determination of thermophysical properties of Si/SiGe superlattices with a pump-probe technique. In Proceedings of the International Workshop on Thermal Investigation of ICs and Systems, Belgirate, Italy, 27–30 September 2005.

48. Chen, G.; Hui, P. Pulsed photothermal modeling of composite samples based on transmission-line theory of heat conduction. *Thin Solid Films* **1999**, *339*, 58–67. [CrossRef]

49. Huang, Z.X.; Tang, Z.A.; Yu, J.; Bai, S. Thermal conductivity of nanoscale polycrystalline ZnO thin films. *Physica B* **2011**, *406*, 818–823. [CrossRef]

50. Ciesa, F.; Plech, A. Gold nanoparticle membranes as large-area surface monolayers. *J. Colloid Interface Sci.* **2010**, *346*, 1–7. [CrossRef] [PubMed]

51. Touloukian, Y.S.; Kirby, R.K.; Taylor, R.E.; Desai, P.D. *Thermal Expansion—Metallic Elements and Alloys*; IFI Plenum: New York, NY, USA, 1975; Volume 12.

52. Suh, I.K.; Ohta, H.; Waseda, Y. High-temperature thermal expansion of six metallic elements measured by dilatation method and X-ray diffraction. *J. Mater. Sci.* **1988**, *23*, 757–760. [CrossRef]

53. Kotaidis, V.; Dekorsy, T.; Ibrahimkutty, S.; Issenmann, D.; Khakhulin, D.; Plech, A. Vibrational symmetry breaking of supported nanospheres. *Phys. Rev. B* **2012**, *86*, 100101. [CrossRef]

54. Zoo, Y.; Adams, D.; Mayer, J.; Alford, T. Investigation of coefficient of thermal expansion of silver thin film on different substrates using X-ray diffraction. *Thin Solid Films* **2006**, *513*, 170–174. [CrossRef]

55. Burzo, M.G.; Komarov, P.L.; Raad, P.E. Thermal Transport Properties of Gold-Covered Thin-Film Silicon Dioxide. *IEEE Trans. Compon. Packag. Technol.* **2003**, *26*, 80–88. [CrossRef]

56. Abadias, G.; Chason, E.; Keckes, J.; Sebastiani, M.; Thompson, G.B.; Barthel, E.; Doll, G.L.; Murray, C.E.; Stoessel, C.H.; Martinu, L. Review Article: Stress in thin films and coatings: Current status, challenges, and prospects featured. *J. Vac. Soc. Technol. A* **2018**, *36*, 020801. [CrossRef]

57. Thomsen, C.; Grahn, H.T.; Maris, H.J.; Tauc, J. Surface generation and detection of phonons by picosecond light pulses. *Phys. Rev. B* **1986**, *34*, 4129. [CrossRef]

58. Schmidt, A.J.; Chen, X.; Chen, G. Pulse accumulation, radial heat conduction, and anisotropic thermal conductivity in pump-probe transient thermoreflectance. *Rev. Sci. Instrum.* **2008**, *79*, 114902. [CrossRef] [PubMed]

59. Frieling, R.; Radek, M.; Eon, S.; Bracht, H.; Wolf, D.E. Phonon coherence in isotopic silicon superlattices. *Appl. Phys. Lett.* **2014**, *105*, 132104. [CrossRef]

60. Frieling, R.; Wolf, D.E.; Bracht, H. Molecular dynamics simulations of thermal transport in isotopically modulated semiconductor nanostructures. *Phys. Status Solidi A* **2016**, *213*, 549–556. [CrossRef]

61. Li, Z.; Tan, S.; Bozorg-Grayeli, E.; Kodama, T.; Asheghi, M.; Delgado, G.; Panzer, M.; Pokrovsky, A.; Wack, D.; Goodson, K.E. Phonon Dominated Heat Conduction Normal to Mo/Si Multilayers with Period below 10 nm. *Nano Lett.* **2012**, *12*, 3121–3126. [CrossRef] [PubMed]

62. Lee, S.M.; Cahill, D.G.; Venkatasubramanian, R. Thermal conductivity of Si-Ge superlattices. *Appl. Phys. Lett.* **1997**, *70*, 2957. [CrossRef]

63. Simkin, M.V.; Mahan, G.D. Minimum Thermal Conductivity of Superlattices. *Phys. Rev. Lett.* **2000**, *84*, 927–930. [CrossRef] [PubMed]

64. Chen, Y.; Li, D.; Lukes, J.R.; Ni, Z.; Chen, M. Minimum superlattice thermal conductivity from molecular dynamics. *Phys. Rev. B* **2005**, *72*, 174302. [CrossRef]

65. Baron, A.Q.R.; Ruby, S.L. Time resolved detection of X-rays using large area avalanche photodiodes. *Nucl. Instrum. Method A* **1994**, *343*, 517–526. [CrossRef]

66. Ibrahimkutty, S.; Issenmann, D.; Schleef, S.; Müller, A.S.; Mathis, Y.L.; Gasharova, B.; Huttel, E.; Steininger, R.; Göttlicher, J.; Baumbach, T.; et al. Asynchronous sampling for ultrafast experiments with low momentum compaction at the ANKA ring. *J. Synchrotron Radiat.* **2011**, *18*, 539–545. [CrossRef]

67. Issenmann, D.; Schleef, S.; Ibrahimkutty, S.; Buth, G.; Baumbach, T.; Beyer, M.; Demsar, J.; Plech, A. Lattice Dynamics of Laser Excited Ferroelectric BaTiO$_3$. *Acta Phys. Pol. A* **2012**, *121*, 319–323. [CrossRef]

68. Schick, D.; von Korff Schmising, C.; Bojahr, A.; Kiel, M.; Gaal, P.; Bargheer, M. Time-Resolved X-Ray Scattering. *Proc. SPIE* **2011**, *7937*, 793715.

nanomaterials

MDPI

Article

On the Formation of Nanocrystalline Grains in Metallic Glasses by Means of In-Situ Nuclear Forward Scattering of Synchrotron Radiation

David Smrčka [1], Vít Procházka [1,*], Vlastimil Vrba [1] and Marcel B. Miglierini [2]

[1] Department of Experimental Physics, Faculty of Science, Palacký University Olomouc, 17. listopadu 12, 771 46 Olomouc, Czech Republic; david.smrcka@upol.cz (D.S.); vlastimil.vrba@upol.cz (V.V.)
[2] Institute of Nuclear and Physical Engineering, Faculty of Electrical Engineering and Information Technology, Slovak University of Technology in Bratislava, Ilkovičova 3, 812 19 Bratislava, Slovakia; marcel.miglierini@stuba.sk
* Correspondence: v.prochazka@upol.cz; Tel.: +420-585-634-302

Received: 28 February 2019; Accepted: 29 March 2019; Published: 4 April 2019

Abstract: Application of the so-called nuclear forward scattering (NFS) of synchrotron radiation is presented for the study of crystallization of metallic glasses. In this process, nanocrystalline alloys are formed. Using NFS, the transformation process can be directly observed during in-situ temperature experiments not only from the structural point of view, i.e., formation of nanocrystalline grains, but one can also observe evolution of the corresponding hyperfine interactions. In doing so, we have revealed the influence of external magnetic field on the crystallization process. The applied magnetic field is not only responsible for an increase of hyperfine magnetic fields within the newly formed nanograins but also the corresponding components in the NFS time spectra are better identified via occurrence of quantum beats with higher frequencies. In order to distinguish between these two effects, simulated and experimental NFS time spectra obtained during in-situ temperature measurements with and without external magnetic field are compared.

Keywords: nuclear forward scattering; metallic glasses; magnetic annealing; synchrotron radiation; crystallization kinetics

1. Introduction

Iron-based metallic glasses (MG) exhibit excellent soft magnetic behaviour because of their high permeability and low coercivity [1,2]. Namely, these properties make them suitable candidates for increasing applications in industry [3] and helping to solve energy-saving problems [4]. Several studies of MGs including the crystallization process, their thermal and magnetic properties [5–8] were reported because they exhibit a wide range of useful physical and structural properties, especially from the application point of view [9,10]. Recently, powder prepared from crushed $Fe_{78}Si_9B_{13}$ amorphous ribbons was used for production of transformer cores with improved magnetic properties obtained by suitable annealing [11] which was eventually performed also in an external magnetic field [12]. The effects of a magnetic field on structural transformations in MGs were reported earlier, for example, in [13]. Nevertheless, the studies performed so far have assessed the effects of an external magnetic field from the point of view of the resulting amount of the crystalline phase and/or their magnetic parameters that were reached after heat treatment. Here, we present a method that can monitor structural transformations in real-time, i.e., during the treatment, namely the so-called nuclear forward scattering (NFS) of synchrotron radiation.

This study aims at a thorough description of transformation processes in iron-based MGs during their conversion into nanocrystalline alloys. Understanding the mechanism of crystallization is important because of succeeding practical applications of these materials as well as from a point of view of basic physical phenomena that are related to structural transformations. To achieve this goal, we must apply novel analytical techniques. Well-established methods comprising X-ray diffraction and differential scanning calorimetry provide information that is averaged over the entire sample and are related exclusively to structural characterization. Recently, the method of NFS, which also scans the magnetic order of the studied systems via their hyperfine interactions, was applied to in-situ investigations of crystallization processes in Fe-Co-Mo-B-Cu MGs [14]. The influence of external magnetic field on the crystallization of FeZrB under isothermal conditions [15] and in FeCuMoB exposed to dynamical temperature increase [16] was studied too.

During crystallization, crystalline grains are formed within the amorphous matrix and, simultaneously, some elements are expelled to the grain boundary regions [15]. Thus, the grains differ from the amorphous matrix both in the long-range order arrangement and in composition. These quantities are reflected in hyperfine parameters which can be inspected by nuclear forward scattering. Formation and development of new crystalline phases in the material can be determined and followed by observing an appearance of new spectral components with different hyperfine parameters.

For example, in iron based amorphous alloys α-Fe nanocrystals develop in the amorphous matrix. Contrary to distribution of quadrupole splitting in the latter, crystalline grains are magnetically ordered, and their hyperfine magnetic fields are close to those of bulk α-Fe [17]. In addition, during the development of nanograins, their hyperfine parameters change too.

Nevertheless, there are cases when the NFS technique experiences difficulties in unambiguous identification of newly formed crystalline phases. This situation occurs when the nanograins are rather small in size ($\sim$10 nm) and their magnetic moments exhibit thermal fluctuations, thus, resulting in an apparent collapse of the hyperfine splitting. In case of iron-based MGs, long-range order in crystalline phases is accompanied by magnetic ordering with rather strong hyperfine magnetic fields (>15 T). In this work, we discuss in detail crystallization processes that occur during the thermal treatment of two MGs with similar compositions, namely $Fe_{57}Co_{20}Mo_8Cu_1B_{14}$ (in this MG Co atoms become part of bcc-Fe(Co) grains) and $Fe_{75}Mo_8Cu_1B_{16}$ in an external magnetic fields of 0.1 T and 0.652 T. Simulation of the influence of an external magnetic field on the corresponding NFS time spectra is also presented.

2. Experimental Details

Metallic glasses of $Fe_{57}Co_{20}Mo_8Cu_1B_{14}$ and $Fe_{75}Mo_8Cu_1B_{16}$ compositions were prepared by planar-flow casting on a rotating quenching wheel. The as-prepared ribbons were 1–2 mm wide and $\sim$20 µm thick. To enhance the count rate, the samples were enriched with the isotope ^{57}Fe to about 50% (note that the natural abundance of ^{57}Fe is 2.17%).

NFS measurements were carried out at the nuclear resonance beamline of the European Synchrotron Radiation Facility (ESRF) in Grenoble. Photon beam energy of 14.413 keV and bandwidth of 3 meV was used to excite the ^{57}Fe nuclei in the samples. Basic principles of the NFS technique are briefly described in [18], more details can be found for example in [19,20].

About 5 mm long pieces of ribbon-shaped samples were placed inside a vacuum furnace installed between two poles of an electromagnet and heated up to 600 °C with a heating rate of 10 °C/min. Experiments were performed without and with an external magnetic field of 0.1 T and 0.652 T. During annealing, each NFS time spectrum was accumulated for one minute. The incident linearly polarized beam entered the sample perpendicularly to its plane and the applied magnetic field was oriented parallel to the polarization axis. The experimental data obtained from NFS measurements were evaluated using the CONUSS software package (version 1.5 by W. Sturhahn, www.nrixs.com) [21,22] in combination with the sequential analysis tool Hubert [23]. During evaluation of the time spectra with the CONUSS software, we took into consideration the transversal coherence length of $\sim$10 µm [24,25], size of the nanograins <15 nm and thickness of the sample $\sim$20 µm (longitudinal coherence length extends far beyond the sample thickness). Under these assumptions, the photons scattered from structurally different regions of the metallic glass, viz. amorphous phase and nanograins, add up coherently.

3. Results and Discussion

3.1. Ferromagnetic $Fe_{57}Co_{20}Mo_8Cu_1B_{14}$ Metallic Glass

Time spectra acquired from the NFS experiments performed upon the $Fe_{57}Co_{20}Mo_8Cu_1B_{14}$ MG are shown in Figure 1. For the sake of more clear presentation of a high number of records, they are plotted as contour plots. The latter are stacked with respect to the duration time of the experiment which constitutes the vertical axes of the contour plots and is directly related to the temperature of annealing. The x-axes represent delayed time which has elapsed between the excitation pulse and the resonantly scattered photons. The counts of the registered photons (intensities) are colour coded in logarithmic scale.

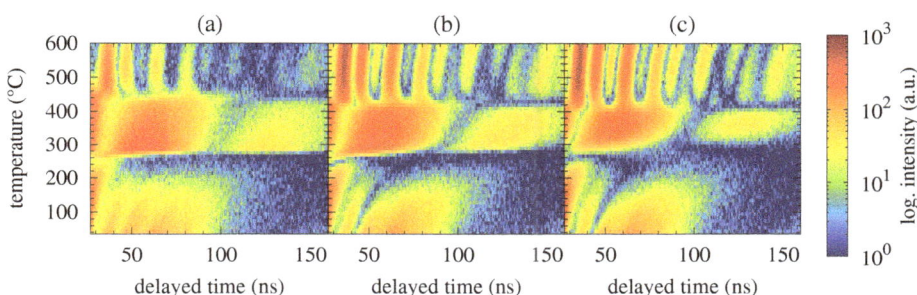

Figure 1. Contour plots of NFS time spectra recorded for the $Fe_{57}Co_{20}Mo_8Cu_1B_{14}$ MG in an external magnetic fields of 0 T (**a**), 0.1 T (**b**) and 0.652 T (**c**).

Two well-distinguished transformations can be seen in Figure 1. The first one is situated at around 250 °C and corresponds to ferromagnetic-to-paramagnetic transition at the Curie temperature. The studied MG is still amorphous and the corresponding beat patterns reflect qualitative change in the respective hyperfine interactions as demonstrated by selected NFS time spectra in Figure 2. Originally weak dipole magnetic interactions, which are characterized by quantum beats with high frequency (temperature 165 °C, zero magnetic field), are in the paramagnetic state of the amorphous alloy replaced by electric quadrupole ones. The latter exhibit rather simple beat patterns, as, for example, those at 375 °C.

The second qualitative change in the NFS time spectra can be seen between 395 °C and 425 °C. Reappearance of quantum beats featuring higher frequencies, which can be first noticed at 405 °C in an external magnetic field of 0.652 T and even more enhanced at a higher temperature of 425 °C, indicates growing importance of dipole magnetic interactions. They mark a presence of hyperfine magnetic fields which belong to ferromagnetic bcc-Fe(Co) nanocrystalline grains, i.e., the onset of crystallization. Chemical composition of this MG ensures that the onset of crystallization is very well visible because it is accompanied by a transition from a paramagnetic amorphous state to a strongly ferromagnetic crystalline state. It is noteworthy that the studied system is still amorphous and paramagnetic at 425 °C when no external magnetic field is applied (Figure 2a).

For the analysis of the NFS time spectra, we have employed a fitting model consisting of five components: one distributed component for the amorphous phase and four components with hyperfine magnetic fields that were ascribed to the signals from bcc-Fe(Co) grains. The number of crystalline components was given by relative probabilities derived from binomial distribution corresponding to 0, 1, 2, and 3 Co atoms as nearest neighbours of Fe. Fitting parameters for the amorphous component comprised relative amount, quadrupole splitting and for all crystalline components relative amount, hyperfine magnetic field, and a parameter related to the effective thickness.

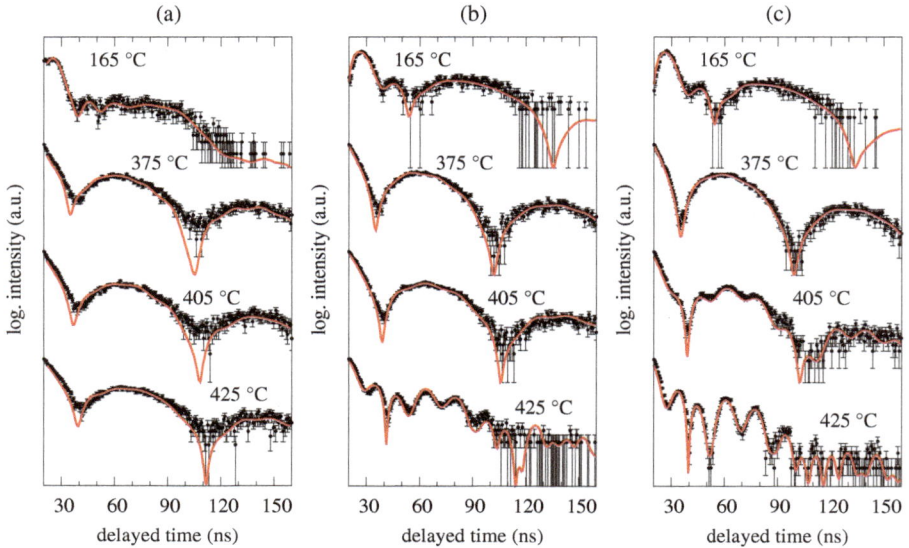

Figure 2. Selected NFS time spectra recorded for the $Fe_{57}Co_{20}Mo_8Cu_1B_{14}$ MG in an external magnetic fields of 0 T (**a**), 0.1 T (**b**) and 0.652 T (**c**) at the indicated temperatures. Black symbols represent experimental data with error margins and red solid lines are results from their fitting.

Temperature evolution of the relative amount of crystalline phase is shown in Figure 3. One can observe an apparent shift of the onset of crystallization towards lower temperatures with increasing external magnetic field. This confirms our assumption that the external magnetic field stimulates the nucleation of grains as it was reported also in other amorphous systems [15,16]. It is noteworthy that the onset of crystallization depends also on the strength of the external magnetic field which was oriented along the length of ribbon-shaped samples.

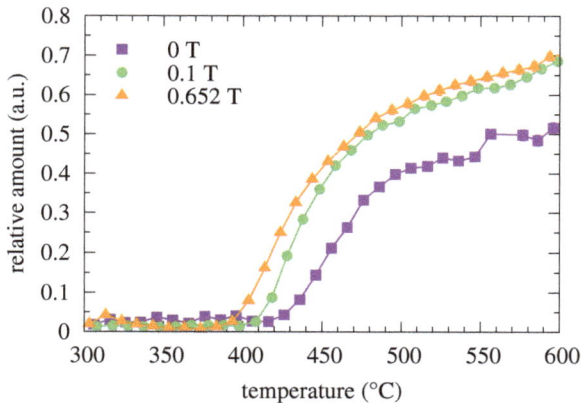

Figure 3. Relative amount of the crystalline phase in the $Fe_{57}Co_{20}Mo_8Cu_1B_{14}$ MG plotted as a function of temperature for different values of external magnetic fields (see the legend).

Hyperfine magnetic fields obtained from the fitting of the crystalline components of the NFS time spectra are shown in Figure 4. They start to appear after the onset of crystallization. With rising temperature of annealing, which also ensures an increase in the relative amount of the crystalline phase (see Figure 3), the hyperfine magnetic fields exhibit clear sharp values. With increasing temperature, the hyperfine magnetic field values follow the expected temperature dependence according to the Brillouin function.

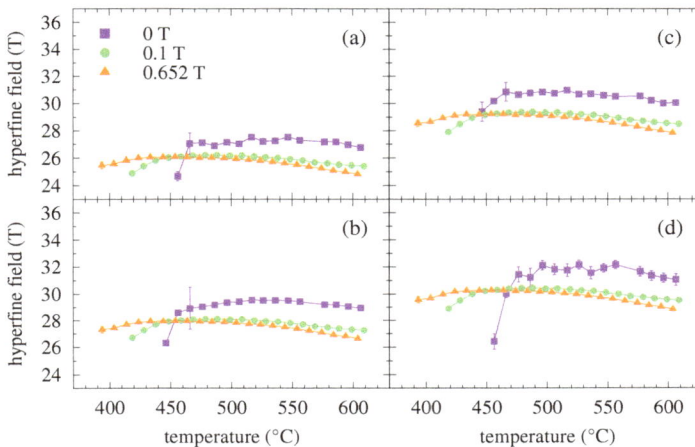

Figure 4. Hyperfine magnetic fields obtained from NFS time spectra of the $Fe_{57}Co_{20}Mo_8Cu_1B_{14}$ MG plotted as a function of temperature for different values of external magnetic fields (see the legend). The plotted hyperfine magnetic fields correspond to the individual crystalline components which represent Fe atoms with 0 (**a**), 1 (**b**), 2 (**c**), and 3 (**d**) Co nearest neighbours.

3.2. Weak Magnetic $Fe_{75}Mo_8Cu_1B_{16}$ Metallic Glass

Unlike the previous case of well-established ferromagnetic $Fe_{57}Co_{20}Mo_8Cu_1B_{14}$ MG where both magnetic and structural transformations are clearly visible from obvious deviations of the NFS beat patterns, the situation is quite different for a weak magnetic $Fe_{75}Mo_8Cu_1B_{16}$ MG. This composition is characterized by a close-to-room Curie temperature of the amorphous precursor ($T_C = 310$ K), small size of bcc-Fe grains (<10 nm) and their low amount (<40%) even towards the end of the primary crystallization [26]. Thus, the onset of crystallization is not accompanied by notable changes in hyperfine interactions because due to its composition, this MG becomes paramagnetic already at the beginning of the heat treatment and the newly emerging Fe nanocrystals are too small and too scarce to ensure that their hyperfine magnetic fields will result in visible higher frequency quantum beats at the onset of crystallization.

Indeed, the acquired NFS time spectra exhibit simple beat patterns which maintain rather uniform structure over a broad temperature range as seen in the corresponding contour plots in Figure 5. The sample remains paramagnetic up to the final annealing temperature. However, as demonstrated by the results of diffraction of synchrotron radiation, this system starts to crystallise at the temperature of ~450 °C [14]. So, even though the bcc-Fe nanocrystals are formed, their presence cannot be confirmed via corresponding hyperfine magnetic fields, which are rather weak, and consequently, the NFS patterns do not show any remarkable changes in their shapes. That is why, alternative ways were proposed how to visualize the presence of nanograins by the help of other parameters derived from NFS time spectra [27].

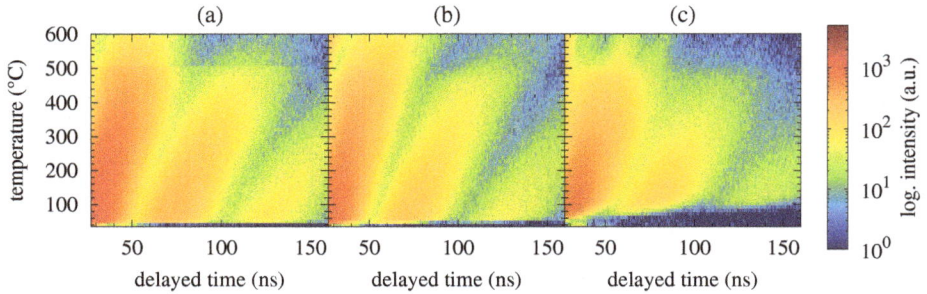

Figure 5. Contour plots of NFS time spectra recorded for the $Fe_{75}Mo_8Cu_1B_{16}$ MG in an external magnetic fields of 0 T (**a**), 0.1 T (**b**) and 0.652 T (**c**).

Examples of experimentally obtained NFS time spectra (black symbols with error bars) together with the fitted curves (solid red lines) are presented in Figure 6 for selected temperatures of annealing. All NFS time spectra were evaluated using a two-component model consisting of one paramagnetic contribution and one component with weak magnetic interactions to refine the amorphous phase. Where necessary, additional narrow magnetic component was used to represent the newly formed crystalline phase. The fitted hyperfine parameters comprised relative amount of each component, quadrupole splitting and hyperfine field for weak magnetic component, hyperfine field for crystalline component and a parameter related to the effective thickness.

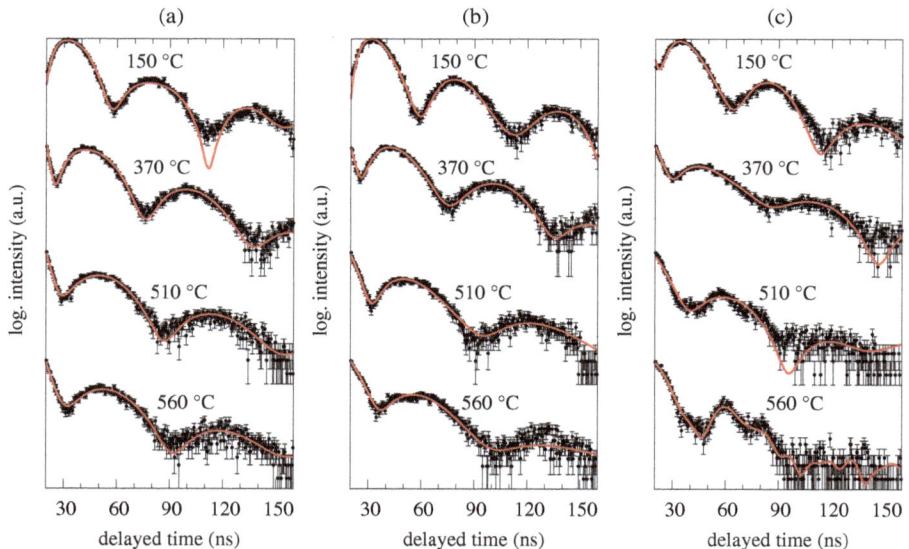

Figure 6. Selected NFS time spectra recorded for the $Fe_{75}Mo_8Cu_1B_{16}$ MG in an external magnetic fields of 0 T (**a**), 0.1 T (**b**) and 0.652 T (**c**) at the indicated temperatures. Black symbols represent experimental data with error margins and red solid lines are results from their fitting.

As already mentioned, all NFS patterns in Figures 5 and 6 show very similar features. Even the application of weak external magnetic field of 0.1 T turned out to have not very pronounced effect and the NFS patterns were practically unchanged. Some deviations in the beat structure can be identified only towards higher temperatures of annealing (>500 °C) in stronger external magnetic field of 0.652 T (Figures 5c and 6c). Here, the newly formed nanocrystalline bcc-Fe grains can be identified via beat patterns with higher frequencies. We assume that the external magnetic field has contributed to their visibility by means of orientation of magnetic moments and consequent increase in magnetization. In this way they had stronger influence on the paramagnetic amorphous phase and, at the same time have formed more uniform magnetic structure of the crystalline phase with apparently stronger average hyperfine magnetic field. The latter is manifested via high-frequency quantum beats in the NFS patterns. Note that in the case of ferromagnetic $Fe_{57}Co_{20}Mo_8Cu_1B_{14}$ MG, even more pronounced high frequency beats can be seen at much lower temperature (~400 °C) and in weak external magnetic field (Figure 2b). Presumably due to magnetic saturation of this soft magnetic metallic glass.

Temperature evolution of the relative amount of the crystalline magnetic phase and its hyperfine magnetic field are shown in Figure 7. They are presented only for the NFS experiment in an external magnetic field of 0.652 T as they were not visible in experiments performed in weaker magnetic fields. The amount of crystalline phase is rather low (<7 %) and the crystalline grains are quite small [26]. Consequently, their magnetic moments fluctuate especially at these high temperatures and are difficult to see through their hyperfine magnetic fields and the corresponding high frequency beat patterns. As far as the hyperfine field values of the crystalline phase are concerned, they follow the expected trend with increasing temperature as demonstrated in Figure 7b.

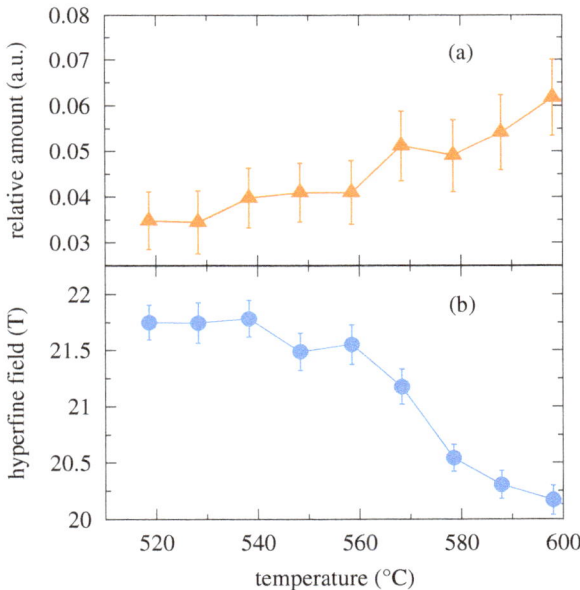

Figure 7. Relative amount of the crystalline phase (**a**) and the corresponding hyperfine magnetic field (**b**) as a function of temperature obtained from the evaluation of NFS time spectra of the $Fe_{75}Mo_8Cu_1B_{16}$ MG performed in an external magnetic field of 0.652 T.

The obtained results suggest that an external magnetic field has an influence on the progress of crystallization especially when magnetic grains are formed. Nevertheless, a question still remains if the external magnetic field affects the process of crystallization as such or if it is only an effect of enhanced visualization of dipole magnetic hyperfine interactions in NFS time spectra especially in weak magnetic MGs. To decide which of these two assumptions is right, we performed simulations of NFS time spectra as presented below.

3.3. Simulations of the Impact of Magnetic Field on NFS Time Spectra

In this section, we describe estimation of the influence of an external magnetic field on a visibility of crystalline components in NFS experiments. Rapid increase of magnetization was reported when the investigated system was placed into an external magnetic field at a temperature close to its Curie temperature [28]. Because hyperfine magnetic field is proportional to magnetization, enhanced Zeeman splitting of nuclear levels occurs when the magnetization increases in the presence of external magnetic field. Consequently, we can observe changes in the shapes of NFS time spectra. Thus, in the $Fe_{75}Mo_8Cu_1B_{16}$ MG, where tiny nanograins are formed, the applied magnetic field triggers formation of magnetic quantum beats with higher frequency. Therefore, two effects can be involved, visualization of magnetic interactions caused by the applied magnetic field as mentioned above and direct influence of the magnetic field on the crystallization process itself.

To distinguish between these two cases, we performed simulations of NFS time spectra. In doing so, the following assumptions were made: (i) hyperfine magnetic field is proportional to the magnetization, (ii) the magnetization of nanocrystals depends on temperature, size of grains, and the applied magnetic field. In applying these assumptions, dependences for α-Fe will be considered including the evolution of Curie temperature with the mean grain size and the dependence of magnetization on temperature and external magnetic field [28,29]. Because in Fe-based MGs, which is also our case, a bcc-Fe crystalline phase is formed, this is a tolerable constraint. Particularly, because it is difficult to obtain values of magnetization from nanograins embedded in a residual amorphous matrix as the magnetic measurements provide integral information from the whole inspected volume.

Taking into account dependencies of magnetization on temperature (T), the applied magnetic field (B_{ext}) and grain size (d) it is possible to construct a function $B_{hf}(T, B_{ext}, d)$ which provides values of hyperfine magnetic field, B_{hf} for arbitrary temperature, applied magnetic field and mean grain size. Because no information on grains size is accessible from NFS experiments, we have used the following procedure. From evaluation of the experimental NFS time spectra for a selected temperature and/or external magnetic field, distribution of hyperfine magnetic fields is readily obtained. The corresponding distribution of grain sizes is calculated by an inverse function $d(T, B_{ext}, B_{hf})$. From that function, we can calculate distribution of hyperfine magnetic fields for any other arbitrarily chosen temperature and/or magnetic field. Subsequently, hypothetical NFS time spectra can be simulated.

We demonstrate the above procedure for the NFS time spectrum that shows in Figure 8a high frequency quantum beats caused by the presence of crystalline grains at 577 °C which have evolved in an external magnetic field of 0.652 T. Using the distribution of hyperfine fields obtained from the fitting, we have simulated NFS time spectrum for the same temperature how it would look like in zero magnetic field. The obtained simulated and the fitted curves are plotted in Figure 8 by blue and red curves, respectively, and compared with the experimental data. Figure 8b shows a situation where the experimental data with the corresponding fitted curve are presented for the same temperature but in zero external magnetic field and overlaid with the simulated spectrum.

It is noteworthy that the blue curves in Figure 8 represent a hypothetical NFS time spectrum as it would look like if the experiment were to be performed in zero magnetic field. It was simulated from the parameters that represent the higher frequency quantum beats from the in-field (0.652 T) experiment. Extrapolating the shapes of the NFS time spectra to zero-field conditions, notable deviations between the fitted experimental spectrum and the simulated one are observed in Figure 8b. This finding, however excludes the hypothesis that external magnetic field only affects the shapes of NFS time spectra thus making the higher frequency quantum beats better visible due to stronger hyperfine magnetic fields of the crystalline phase induced by enhanced magnetization. In other words, the high frequency quantum beats in the simulated NFS spectrum result from magnetic effects upon the formation of nanocrystalline grains. The same conclusion is supported by Figure 8a where deviations between the fitted and the simulated NFS spectrum can be seen, too.

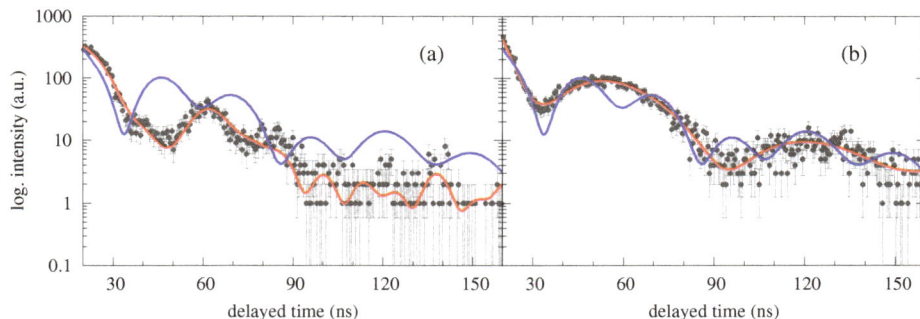

Figure 8. NFS time spectra of the $Fe_{75}Mo_8Cu_1B_{16}$ MG recorded at 577 °C in an external magnetic field of 0.652 T (**a**) and in zero field (**b**). Red solid curves represent results from the fitting, blue curves are simulations for 0 T using the fitted data for 0.652 T (see text).

4. Conclusions

Formation of nanocrystalline grains during in-situ heat treatment was followed by the help of nuclear forward scattering of synchrotron radiation. This method offers on fly information in real-time both on the structural arrangement and associated hyperfine interactions in the studied system. Possibilities of this unique technique were demonstrated using two types of Fe-based metallic glasses featuring opposite magnetic behaviour. Namely, we have used amorphous alloys with the compositions of $Fe_{57}Co_{20}Mo_8Cu_1B_{14}$ and $Fe_{75}Mo_8Cu_1B_{16}$ which exhibit strong magnetic interactions and weak magnetic order, respectively.

Formation of ferromagnetic crystalline grains in the amorphous matrix is accompanied by occurrence of high frequency quantum beats in the NFS time spectra. They reflect the onset of crystallization differently in both investigated systems. While in ferromagnetic $Fe_{57}Co_{20}Mo_8Cu_1B_{14}$ metallic glass the existence of bcc-Fe(Co) nanograins is readily seen, modifications of the shapes of NFS time spectra of a weakly magnetic $Fe_{75}Mo_8Cu_1B_{16}$ metallic glass are almost unnoticeable. All NFS experiments were performed also in an external magnetic fields of 0.1 T and 0.652 T.

For the $Fe_{57}Co_{20}Mo_8Cu_1B_{14}$ metallic glass, a transition from ferromagnetic to paramagnetic state was observed at temperature of ~250 °C which was later followed by a structural transformation from amorphous to nanocrystalline arrangement. Evaluation of experimental NFS data revealed the influence of external magnetic field on the crystallization process. The applied magnetic field shifts the onset of crystallization towards lower temperatures. Simultaneously, it increases the fraction of crystalline grains embedded in the amorphous matrix.

In case of weak magnetic $Fe_{75}Mo_8Cu_1B_{16}$ metallic glass, presence of bcc-Fe nanograins during temperature annealing in zero field and in the weak external magnetic field of 0.1 T is not accompanied by obvious presence of corresponding quantum beats in the NFS time spectra. Therefore, identification of the onset of crystallization is not straightforward. This is mainly due to formation of tiny nanocrystalline grains whose magnetic moments significantly fluctuate especially at high enough temperatures. In the applied magnetic field of 0.652 T, we observed formation of nanocrystalline grains with sufficiently well-developed magnetic interactions.

To assess the influence of an external magnetic field upon the NFS experiments, we have performed simulations of the time spectra. The obtained results allow a conclusion that the accelerated crystallization, which was observed under the effect of external magnetic field in several metallic glasses investigated so far, is caused by magnetic energy from this field. The appearance of high frequency quantum beats in the shapes of NFS time spectra is not purely an artefact but a real demonstration of a presence of crystalline phase.

Author Contributions: Conceptualization, M.B.M. and V.P.; Methodology, D.S. and V.V.; Formal analysis, D.S.; Validation, V.P. and V.V.; Visualization, D.S.; Writing—Original Draft Preparation, V.P. and M.B.M.

Funding: This research was funded by grant VEGA 1/0182/16/ and the internal IGA grant of Palacký University (IGA_PrF_2018_002).

Acknowledgments: We are grateful to R. Rüffer for his assistance during experiments at nuclear resonant beamline at ESRF in Grenoble.

Conflicts of Interest: The authors declare no conflict of interest.

Abbreviations

The following abbreviations are used in this manuscript:

MG	Metallic glass
NFS	Nuclear Forward Scattering
ESRF	European Synchrotron Radiation Facility

References

1. Chang, Y.H.; Hsu, C.H.; Chu, H.L.; Chang, C.W.; Chan, W.S.; Lee, C.Y.; Yao, C.S.; He, Y.L. Effect of uneven surface on magnetic properties of Fe-based amorphous transformer. *World Acad. Sci. Eng. Technol.* **2011**, *56*, 1435–1439.
2. Herzer, G.; Hilzinger, H. Surface crystallization and magnetic properties in amorphous iron rich alloys. *J. Magn. Magn. Mater.* **1986**, *62*, 143–151. [CrossRef]
3. Wu, C. Development of amorphous ribbon manufacturing technology. *China Steel Tech. Rep.* **2014**, *27*, 28–42.
4. Hasegawa, R. Present status of amorphous soft magnetic alloys. *J. Magn. Magn. Mater.* **2000**, *215*, 240–245. [CrossRef]
5. Zhao, L.Z.; Yu, H.Y.; Li, W.; Liao, X.F.; Zhang, J.S.; Zhong, X.C.; Liu, Z.W.; Su, K.P.; Greneche, J.M. Magnetic characteristics of the ferromagnetic Fe-rich clusters in bulk amorphous $Nd_{60}Fe_{30}Al_{10}$ alloy. *J. Magn. Magn. Mater.* **2019**, *469*, 151–154. [CrossRef]
6. Parsons, R.; Zang, B.; Onodera, K.; Kishimoto, H.; Kato, A.; Suzuki, K. Soft magnetic properties of rapidly-annealed nanocrystalline Fe-Nb-B-(Cu) alloys. *J. Alloy. Compd.* **2017**, *723*, 408–417. [CrossRef]
7. Salazar, D.; Martín-Cid, A.; Madugundo, R.; Garitaonandia, J.S.; Barandiaran, J.M.; Hadjipanayis, G.C. Effect of Nb and Cu on the crystallization behavior of under-stoichiometric Nd–Fe–B alloys. *J. Phys. D Appl. Phys.* **2016**, *50*, 015305. [CrossRef]
8. Barandiarán, J.M.; Gutiérrez, J.; García-Arribas, A. Magneto-elasticity in amorphous ferromagnets: Basic principles and applications. *Phys. Status Solidi (A)* **2011**, *208*, 2258–2264. [CrossRef]
9. Gutiérrez, J.; Lasheras, A.; Martins, P.; Pereira, N.; Barandiarán, J.M.; Lanceros-Mendez, S. Metallic Glass/PVDF Magnetoelectric Laminates for Resonant Sensors and Actuators: A Review. *Sensors* **2017**, *17*, 1251. [CrossRef] [PubMed]
10. Zhang, Y.; Ramanujan, R. A study of the crystallization behavior of an amorphous $Fe_{77.5}Si_{13.5}B_9$ alloy. *Mater. Sci. Eng. A* **2006**, *416*, 161–168. [CrossRef]
11. Li, Z.; Dong, Y.; Li, F.; Chang, C.; Wang, X.M.; Li, R.W. $Fe_{78}Si_9B_{13}$ amorphous powder core with improved magnetic properties. *J. Mater. Sci.-Mater. Electron.* **2017**, *28*, 1180–1185. [CrossRef]
12. Li, Z.; Dong, Y.; Pauly, S.; Chang, C.; Wei, R.; Li, F.; Wang, X.M. Enhanced soft magnetic properties of Fe-based amorphous powder cores by longitude magnetic field annealing. *J. Alloy. Compd.* **2017**, *706*, 1–6. [CrossRef]
13. Yardley, V.A.; Tsurekawa, S.; Fujii, H.; Matsuzaki, T. Thermodynamic Study of Magnetic Field-Enhanced Nanocrystallisation in Amorphous Fe-Si-B(-Nb-Cu). *Mater. Trans.* **2007**, *48*, 2826–2832. [CrossRef]
14. Miglierini, M.; Pavlovič, M.; Procházka, V.; Hatala, T.; Schumacher, G.; Rüffer, R. Evolution of structure and local magnetic fields during crystallization of HITPERM glassy alloys studied by in situ diffraction and nuclear forward scattering of synchrotron radiation. *Phys. Chem. Chem. Phys.* **2015**, *17*, 28239–28249. [CrossRef]
15. Miglierini, M.; Procházka, V.; Rüffer, R.; Zbořil, R. *In situ* crystallization of metallic glasses during magnetic field annealing. *Acta Mater.* **2015**, *91*, 50–56. [CrossRef]

16. Procházka, V.; Vrba, V.; Smrčka, D.; Rüffer, R.; Matúš, P.; Mašláň, M.; Miglierini, M.B. Structural transformation of NANOPERM-type metallic glasses followed in situ by synchrotron radiation during thermal annealing in an external magnetic field. *J. Alloys Compd.* **2015**, *638*, 398–404. [CrossRef]

17. Preston, R.; Heberle, J.; Hanna, S. Mössbauer Effect in Metallic Iron. *Phys. Rev.* **1962**, *128*, 2207–2218. [CrossRef]

18. Miglierini, M.B.; Procházka, V. Nanocrystallization of Metallic Glasses Followed by in situ Nuclear Forward Scattering of Synchrotron Radiation. In *X-ray Characterization of Nanomaterials by Synchrotron Radiation*; InTech: London, UK, 2017; pp. 7–29.

19. Röhlsberger, R. *Nuclear Condensed Matter Physics with Synchrotron Radiation*; Springer: Berlin/Heidelberg, Germany, 2005; Volume 208, p. 318.

20. Rüffer, R. Nuclear resonance scattering. *Comptes Rendus Phys.* **2008**, *9*, 595–607. [CrossRef]

21. Sturhahn, W.; Gerdau, E. Evaluation of time-differential measurements of nuclear-resonance scattering of x rays. *Phys. Rev. B* **1994**, *49*, 9285–9294. [CrossRef]

22. Sturhahn, W. CONUSS and PHOENIX: Evaluation of nuclear resonant scattering data. *Hyperfine Interact.* **2000**, *125*, 149–172. [CrossRef]

23. Vrba, V.; Procházka, V.; Smrčka, D.; Miglierini, M. Advanced approach to the analysis of a series of in-situ nuclear forward scattering experiments. *Nucl. Instrum. Methods Phys. Res. Sect. A Accel. Spectrom. Detect. Assoc. Equip.* **2017**, *847*, 111–116. [CrossRef]

24. Baron, A.Q.R.; Chumakov, A.I.; Grünsteudel, H.F.; Grünsteudel, H.; Niesen, L.; Rüffer, R. Transverse X-Ray Coherence in Nuclear Scattering of Synchrotron Radiation. *Phys. Rev. Lett.* **1996**, *77*, 4808–4811. [CrossRef] [PubMed]

25. Baron, A.Q.R. Transverse coherence in nuclear resonant scattering of synchrotron radiation. *Hyperfine Interact.* **1999**, *123*, 667–680. [CrossRef]

26. Paluga, M.; Švec, P.; Janičkovič, D.; Muller, D.; Mrafko, P.; Miglierini, M. Nanocrystallization in rapidly quenched Fe-Mo-Cu-B: Surface and volume effects. *Rev. Adv. Mater. Sci.* **2008**, *18*, 481–493.

27. Smrčka, D.; Procházka, V.; Vrba, V.; Miglierini, M. Nuclear forward scattering analysis of crystallization processes in weakly magnetic metallic glasses. *J. Alloys Compd.* under review.

28. Craig, P.; Perisho, R.; Segnan, R.; Steyert, W. Temperature and field dependence of hyperfine fields and magnetization in a dilute random substitutional ferromagnetic alloy: $Fe_{2.65}Pd_{97.35}$. *Phys. Rev.* **1965**, *138*, 1460–1471. [CrossRef]

29. Delavari, H.; Madaah Hosseini, H.; Simchi, A. A simple model for the size and shape dependent Curie temperature of freestanding Ni and Fe nanoparticles based on the average coordination number and atomic cohesive energy. *Chem. Phys.* **2011**, *383*, 1–5. [CrossRef]

nanomaterials

MDPI

Article

Direct Spectroscopy for Probing the Critical Role of Partial Covalency in Oxygen Reduction Reaction for Cobalt-Manganese Spinel Oxides

Xinghui Long [1,2,3], Pengfei Yu [1,2], Nian Zhang [1,2], Chun Li [4], Xuefei Feng [5], Guoxi Ren [1,2,3], Shun Zheng [1,2,3], Jiamin Fu [1,2,6], Fangyi Cheng [4] and Xiaosong Liu [1,2,6,*]

[1] State Key Laboratory of Functional Materials for Informatics, Shanghai Institute of Microsystem and Information Technology, Chinese Academy of Sciences, Shanghai 200050, China; xhlong@mail.sim.ac.cn (X.L.); ypfaq@mail.sim.ac.cn (P.Y.); zhangn@mail.sim.ac.cn (N.Z.); gxren@mail.sim.ac.cn (G.R.); shunzheng@mail.sim.ac.cn (S.Z.); fujm@shanghaitech.edu.cn (J.F.)
[2] CAS Center for Excellence in Superconducting Electronics (CENSE), Chinese Academy of Sciences, Shanghai 200050, China
[3] University of Chinese Academy of Sciences, Beijing 100049, China
[4] Key Laboratory of Advanced Energy Materials Chemistry (Ministry of Education) and State Key Laboratory of Elemento-Organic Chemistry, College of Chemistry, Nankai University, Tianjin 300071, China; kemistlic@foxmail.com (C.L.); fycheng@nankai.edu.cn (F.C.)
[5] Advanced Light Source, Lawrence Berkeley National Laboratory, Berkeley, CA 94720, USA; xuefeifeng2013@gmail.com
[6] School of Physical Science and Technology, Shanghai Tech University, Shanghai 200031, China
* Correspondence: xliu3@mail.sim.ac.cn; Tel.: +86-021-6251-1070

Received: 25 February 2019; Accepted: 1 April 2019; Published: 9 April 2019

Abstract: Nanocrystalline multivalent metal spinels are considered as attractive non-precious oxygen electrocatalysts. Identifying their active sites and understanding their reaction mechanisms are essential to explore novel transition metal (TM) oxides catalysts and further promote their catalytic efficiency. Here we report a systematic investigation, by means of soft X-ray absorption spectroscopy (sXAS), on cubic and tetragonal $Co_xMn_{3-x}O_4$ (x = 1, 1.5, 2) spinel oxides as a family of highly active catalysts for the oxygen reduction reaction (ORR). We demonstrate that the ORR activity for oxide catalysts primarily correlates to the partial covalency of between O 2p orbital with Mn^{4+} 3d t_{2g}-down/e_g-up, Mn^{3+} 3d e_g-up and Co^{3+} 3d e_g-up orbitals in octahedron, which is directly revealed by the O K-edge sXAS. Our findings propose the critical influences of the partial covalency between oxygen 2p band and specific metal 3d band on the competition between intermediates displacement of the ORR, and thus highlight the importance of electronic structure in controlling oxide catalytic activity.

Keywords: oxygen reduction reaction; spinel oxides; soft X-ray absorption spectroscopy; partial covalency; catalytic activity

---

1. Introduction

The oxygen reduction reaction (ORR) and/or oxygen evolution reaction (OER) on an oxygen-based electrode are essential for a wide range of electrochemical energy conversion and storage technologies, such as direct solar cell [1], electrolytic water splitting [2], rechargeable metal–air batteries [3], and regenerative fuel cells [4]. However, the intrinsic slow kinetics of ORR/OER is the obstacle for their application. It is a great challenge to seek for highly active catalysts to improve the efficiency of ORR and OER. To date, the best known catalysts for oxygen electrocatalysis are Pt-alloy catalysts for the ORR [5] and iridium-oxide- or ruthenium-oxide-based catalysts for the

OER [6]. Unfortunately, the scarce crustal abundance of the noble metals limits their commercial viability. Transition metal (TM) oxides and carbon materials with excellent electrocatalysts and high stability [7–9], owing many advantages such as high abundance, low-cost, easy prepared, and environmental friendliness, are considered as an alternative to noble metals. In particularly, spinel oxides have been widely used as catalyst for ORR and/or OER [10–16]. In pursuit of further enhanced oxygen electrocatalytic activity, it is necessary to understand the catalytic mechanism of TM oxides and to identify the activity site, which has attracted extensive research efforts.

For instance, Rios et al. found that the surface Co^{3+} in $Mn_xCo_{3-x}O_4$ ($x > 0$) was the active site, which made Co_3O_4 the most active for OER [17]. On the contrary, Restovic et al. pointed out that the electrocatalytic activity in $Mn_xCo_{3-x}O_4$ of the ORR was correlated to the Mn content, and more precisely to the amount of Mn^{4+}/Mn^{3+} pairs [18]. These results suggested that two metals in dual-metal spinel system and their redox pairs play different roles in influencing the catalytic performances for ORR or OER. Based on a systematic study of 3d TM perovskite oxides, Shao-Horn Yang et al. discovered a volcano shape relationship between the e_g-filling descriptor, depicting filling degree of the surface active ions e_g orbital, and the catalytic activity of ORR/OER [19,20]. Almost at the same time, Zhichuan J. Xu et al. speculated that the e_g occupancy of the active cation in the octahedral site is the activity descriptor for the ORR/OER of spinels [21,22]. They also elucidated, based on an investigation of the composition dependence of ORR in $ZnCo_xMn_{2-x}O_4$ ($x = 0.0–2.0$) spinel, that the modulated e_g occupancy of active Mn cations, as a consequence of the superexchange effect between edge sharing [CoO_6] and [MnO_6] octahedra, correlated to the ORR activity [22]. However, David N. Mueller et al. revealed that the oxygen anions near the surface rather than the TM cations were a significant redox partner to molecular oxygen because of the strong covalency between oxygen 2p orbital and TM 3d orbital in oxygen-deficient perovskite oxides [23]. In addition, the covalency between metal-d and oxygen-p had been reported to play a critical role in increasing the activities of oxygen electrocatalysis [19,24]. Obviously, although the spinel oxides have been extensively studied, it is still under debate about their catalytic mechanism of ORR or OER activity, especially for dual-metal spinel oxides.

To tackle this long-standing and important fundamental problem, we herein report a systematic and detailed study of a series of nanocrystalline dual-metal spinals $Co_xMn_{3-x}O_4$ through a detailed study with soft X-ray absorption spectroscopy (sXAS). The aforementioned facile synthesis methodology [25] facilitates selective formation of cubic or tetragonal phases and various compositions of $Co_xMn_{3-x}O_4$ ($x = 1, 1.5, 2$) nanoparticles, which are believed as two main factors affecting their ORR catalytic activities. The subtle variation in TM L-edge spectra for both cubic and tetrahedral samples indicates very little valance change of TM with the variation of the component proportion. Surprisingly, the notable differences are observed, associated with the substitution of Co by Mn, from O K-edge absorption spectra, in particular the pre-edge structures that arise from the covalent mixture of metal-3d and oxygen-2p electronic states. These covalent characteristics is analyzed in depth through the deconvolution of pre-edge features and comparison to a series of reference samples to identify their origins. Our results suggest that the partial covalency of O 2p orbital with Mn^{4+} 3d t_{2g}-down/e_g-up, Mn^{3+} 3d e_g-up and Co^{3+} 3d e_g-up orbitals have stronger correlation than other orbitals to the ORR catalytic activities. These findings may provide a new experimental evidence from the point of view of electronic structure to unveil the catalytic mechanism of dual-metal spinels.

2. Experimental Method

2.1. Synthesis of $Co_xMn_{3-x}O_4$ Oxides and Electrochemical Characterization

The $Co_xMn_{3-x}O_4$ oxides were obtained by a solution synthesis method. In a typical synthesis of $Co_xMn_{3-x}O_4$ spinel oxide, $Co(NO_3)_2$ and $Mn(NO_3)_2$ (Sigma Aldrich, St. Louis, MO, USA) were used as precursor of cobalt and manganese, respectively. The specific steps can be summarized as stirring the solution containing aqueous ammonia (Sigma Aldrich, St. Louis, MO, USA), $Co(NO_3)_2$ and $Mn(NO_3)_2$

solution, then evaporating by heating to obtain the final spinel oxide. Different ratios x and phase structure were obtained by controlling the molar ratio of cobalt and manganese precursor and the order of adding ammonia water, $Co(NO_3)_2$ and $Mn(NO_3)_2$ solution, respectively.

To test the electrochemical performance, a three electrodes electrochemical cell were used, which contains a calomel reference electrode, a Pt counter electrode and a working electrode, respectively. Catalyst ink, containing spinels oxide, carbon powder, water, isopropyl alcohol and neutralized Nafion solution (Sigma Aldrich, St. Louis, MO, USA), was pipetted on the glassy carbon electrode to form the working electrode. The PARSTAT 263A workstation (AMETEK, Berwyn, PA, USA) accompanied with a model 636 system (AMETEK, Berwyn, PA, USA) was used to record the voltammetry data with a potential scan rate of 5 mV s^{-1}. Measurements were carried out in 0.1 M aqueous KOH saturated with either purified Ar or O_2 at room temperature. All potentials were calibrated with reference to standard reversible hydrogen electrode. The detailed experimental methods can refer to reference [26].

2.2. Soft X-ray Absorption Spectroscopy (sXAS)

sXAS were performed at beamline 20A1 of National Synchrotron Radiation Research Center (NSRRC) in Hsinchu, Taiwan. The storage ring was operated with energy of 1.5 GeV and a current of 300 mA. The beamline was equipped with a 6-m high-energy spherical grating monochromator (6m-HSGM) to supply a photon beam with resolving power up to 8000 [27]. The spectra were collected in total electron yield (TEY) mode in an under ultrahigh-vacuum (UHV) chamber with a base pressure about 5×10^{-10} Torr, corresponding to probe depth of about 10 nm. All the spectra have been normalized to the photocurrent from the upstream clean gold mesh to eliminate the fluctuation of the beam flux. The photon energy was calibrated with the spectra of reference samples (MnO for Mn L-edge, CoO for Co L-edge and $SrTiO_3$ for O K-edge) measured simultaneously.

3. Results and Discussion

Cubic and tetragonal spinel $Co_xMn_{3-x}O_4$ (x = 1, 1.5, 2) samples (labeled as C–Co$_1$, C–Co$_{1.5}$, C–Co$_2$, T–Co$_1$, T–Co$_{1.5}$, and T–Co$_2$, respectively) are obtained by solution synthesis method and their crystal structure are shown in Figure 1a,c. Spinel oxides have A[B$_2$]X$_4$ molecular formula, in which A stands for cations occupied eighth of the tetrahedral sites, B stands for cations occupied half of the octahedral sites, and X represents oxygen anions with a close-packed structure. The structure, morphology and ORR performance of $Co_xMn_{3-x}O_4$ (x = 1, 1.5, 2) can be found in the previous report [26]. The different ORR performance can be viewed from the half-wave potential E$_{1/2}$ versus reversible hydrogen electrode (RHE) acquired from the polarization profiles, as shown in Figure 1b,d. For both cubic and tetrahedral phases, the activities decrease with the increase of Co/Mn ratio, i.e., CoMn$_2$ > Co$_{1.5}$Mn$_{1.5}$ > Co$_2$Mn. The results indicate that lower Co/Mn ratio are more favorable to intrinsic catalytic activity.

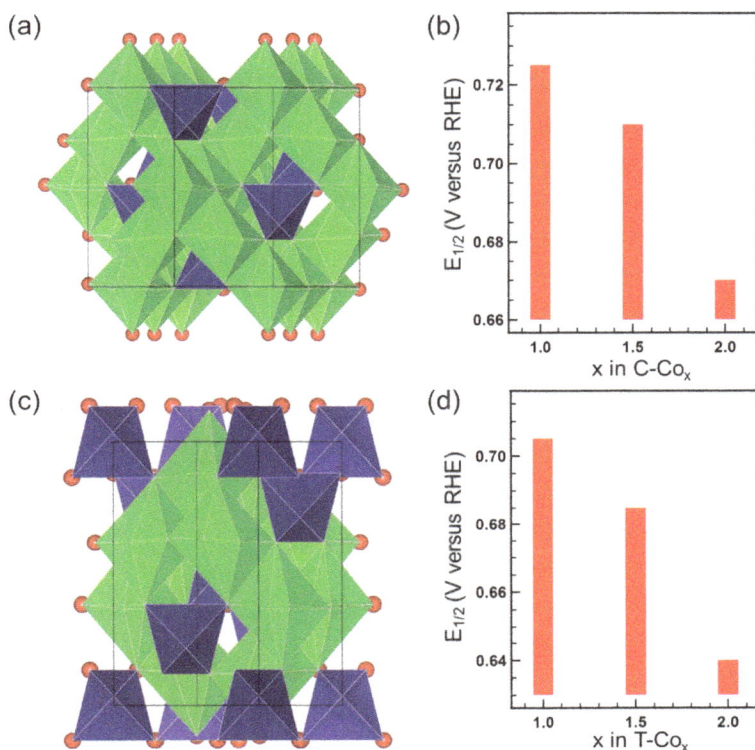

Figure 1. (**a**,**c**) Structure diagram of cubic and tetragonal spinel oxides, respectively. The red ball, blue polyhedron, and green polyhedron represent the oxygen anion, tetrahedron, and octahedron, respectively. (**b**,**d**) The half-wave potential $E_{1/2}$ versus the Co content x in the cubic and tetragonal series, respectively. (5 mV s^{-1} scan rate).

To reveal the effect of the Co/Mn ratio on the electronic structure and get a deep insight into its relationship with the catalytic performance, TM L-edge sXAS spectra are firstly studied. The sXAS spectra of Co and Mn L-edge of cubic and tetragonal $Co_xMn_{3-x}O_4$ (x = 1, 1.5, 2) under UHV condition are shown in Figure 2a,b, respectively. It can be seen that both the Co and Mn L-edge spectra split into two separate sets of peaks named as L_3 and L_2-edge as a result of the 2p spin-orbital coupling interaction. We herein focus on the evolution of L_3-edge because of the stronger intensities and refined features compared to L_2-edge. Based on the reference samples with different valence states and published results [28–31], the peak A (~779 eV) and peak B (~780.6 eV) are assigned to the Co^{2+} and Co^{3+} states, respectively. The overall lineshape indicates that the Co of $Co_xMn_{3-x}O_4$ primarily exist in a mixed 2+/3+ oxidation state. The subtle spectral variations with different Co/Mn ratios imply little valance changes of Co for both the cubic and tetragonal phases. Moreover, all spectra of $Co_xMn_{3-x}O_4$ are very similar to the spectrum of spinel Co_3O_4, suggesting that the Co^{2+} and Co^{3+} occupy the tetrahedral and octahedral sites, respectively [28,32–34]. The Mn L-edge sXAS spectra show similar phenomena. Following the same analysis, we are able to identify that the oxidations of Mn in $Co_xMn_{3-x}O_4$ primarily exist in a mixed 3+/4+ state in comparison with previous studies [29,35–37]. The similarity of spectra between $Co_xMn_{3-x}O_4$ and spinel $LiMn_2O_4$ reveals that both Mn^{3+} and Mn^{4+} occupy the octahedral sites. More distinct variation with the alteration of x in $Co_xMn_{3-x}O_4$ are observed in the Mn L-edge spectra than the Co L-edge. As the valance state of Mn is higher than Co, the valance changes state that TM or O vacancies are created during the synthetic process, which is consistent with previous report [26]. Besides, it has been well established that the high spin states correspond to large branching ratio $I(L_3)/[I(L_3) + I(L_2)]$ [38,39]. The small branching ratio at Co L-edge and large in Mn L-edge clear that Co stays at low-spin state [31] and Mn at high-spin state [40] in all samples, respectively.

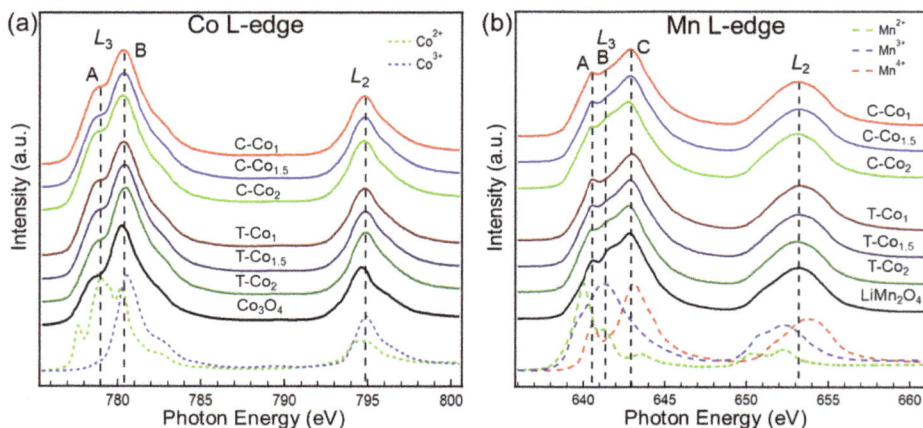

Figure 2. (a) Co L-edge soft X-ray absorption spectroscopy (sXAS) spectra of the spinel oxides with Co^{2+} (CoO), Co^{3+} (LiCoO$_2$) and Co$_3$O$_4$. (b) Mn L-edge sXAS spectra of spinel oxides with Mn^{2+} (MnO), Mn^{3+} (Mn$_2$O$_3$), Mn^{4+} (Li$_2$MnO$_3$), and LiMn$_2$O$_4$.

Since very little changes from TM L-edge are observed with the alteration of Co/Mn ratio, O K-edge is examined to further track the possible evolution of electronic structure. The O K-edge sXAS spectra of the spinel oxides in Figure 3a can be divided into two regions. The first region (529 ~ 535 eV) shown in the shade, so-called pre-edge, is primarily associated with the O 1s to unoccupied O 2p-TM 3d hybridized states. The second region (above 535 eV) has been attributed to the excitations of O 1s to O 2p−TM 4sp states [23,24,29,41]. Different from the TM L-edge, a significant change is observed in the O 2p−TM 3d region of O K-edge. The energy position of the peak A shifts to high energy as the cobalt content x increases, and the intensity ratio between peak A and B also changes for both cubic and tetragonal phases. This suggests that the M 3d-O 2p covalency may be regulated by changing the ratio x in Co$_x$Mn$_{3-x}$O$_4$ oxides. In order to further testify if there exists an ORR activity-determining factor related to M 3d-O 2p covalency, we investigate how the ORR activity and M 3d-O 2p covalency changes as a function of x in Co$_x$Mn$_{3-x}$O$_4$. The normalized absorbance percentage, which is estimated by the percentage of shaded area with subtracting a linear background relative to the entire area of the curve showed in Figure 3a, is used to quantify the strength of M 3d-O 2p covalency. To identify the effect of the covalency of M 3d-O 2p on the ORR activity, we plot the normalized absorbance percentage versus the half-wave potential $E_{1/2}$ in Figure 3b. The M 3d-O 2p covalency exhibits a consistent variation trend with the ORR performance in the tetragonal phase, which demonstrates that increasing the M 3d-O 2p covalency positively affects ORR activity. On the contrary, this correlation cannot be observed in the cubic phase. It evidenced that the strength of the M 3d-O 2p covalency is not a common descriptor of the ORR performance for both the cubic and tetragonal phases. More complicated underlying mechanism may play more crucial role to determine their catalytic activity.

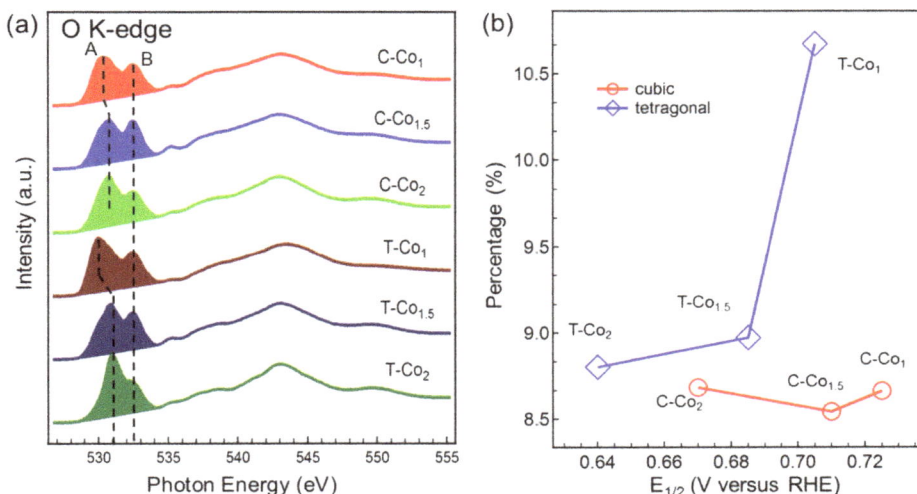

Figure 3. The role of M 3d-O 2p covalency on the oxygen reduction reaction (ORR) activity of $Co_xMn_{3-x}O_4$ spinel oxides. (**a**) O K-edge sXAS spectra of cubic and tetragonal $Co_xMn_{3-x}O_4$ (x = 1, 1.5, 2). Reproduced with permission from [26]. Copyright Nature Publishing Group, 2015. (**b**) the normalized absorbance percentage (absorbance percentage from the shaded section in A) versus the half-wave potential $E_{1/2}$.

The complexity of dual-metal spinels and the abundant features in the pre-edge of O K-edge sXAS spectra inspire us to perform a more in-depth analysis. First of all, we compare a series of reference samples (Co_3O_4, $LiCoO_2$, Li_2MnO_3, Mn_2O_3) along with our sample C–Co_1, as shown in the Figure 4a, to figure out the contributions of the Co^{2+} at tetrahedral site as well as the Co^{3+}, Mn^{3+}, and Mn^{4+} at octahedral sites to O K-edge spectra. The pre-edge features of Co_3O_4 are very similar to the $LiCoO_2$, except for an additional weak shoulder at ~532.6 eV corresponding to the Co^{2+} at tetrahedral site. This declares that the contribution of Co^{2+} in the tetrahedron to the O–K pre-edge is negligible in spinel oxides. For Mn-containing oxides, two well-resolved peaks are observed in Li_2MnO_3 and Mn_2O_3 and their lineshapes are very different. The high-energy peak of Mn_2O_3 is broadening and probably the superposition of two peaks [36]. A glancing comparison between the C–Co_1 and these reference samples illustrates, as shown by the dotted line in the Figure 4a, that the pre-edge of $Co_xMn_{3-x}O_4$ contains all specific features, which can be considered as the spectral fingerprint of Co^{2+} at tetrahedral site as well as the Co^{3+}, Mn^{3+}, and Mn^{4+} at octahedral sites. Furthermore, we perform a quantitative analysis by spectral fitting method, as shown in Figure 4b (and Figure S1 in Supplementary Materials) [42,43]. In detail, symmetrically constrained Gaussian features and an arctangent function background are employed to do curve fitting. The full width at half maximum (FWHM) and energy position of the Gaussian functions and the arctangent background for the spectra peak deconvolution are listed in Tables S1 and S2 in supplementary materials. The most important information gained from this analysis is that the O–K pre-edge of $Co_xMn_{3-x}O_4$ can be deconvoluted by four well-resolved intense Gaussian features labeled as P1–P4, representing four partial covalency of different Co–O and Mn–O. In addition, their intensities can be determined by the area of the corresponding curve-fitting functions.

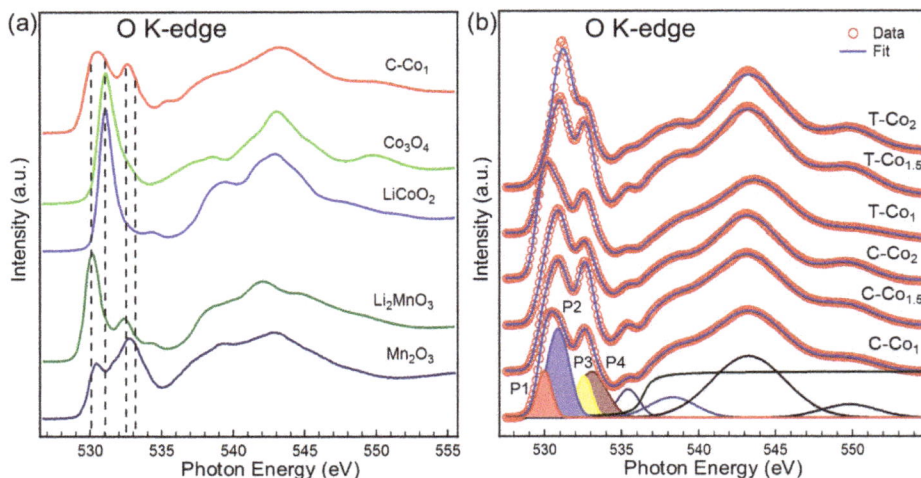

Figure 4. (**a**) O K-edge sXAS spectra of Mn_2O_3, Li_2MnO_3, $LiCoO_2$, Co_3O_4, and C-Co$_1$. (**b**) O K-edge XAS signals of the spinel oxides; the peak decomposition has also been indicated in the figure. The experimental data are shown with the open red circles and the fitted results are shown with the solid blue line. Shaded Gaussian peaks (P1, P2, P3, and P4) represent the M 3d-O 2p covalency.

To further visualize the correlations between the P1–P4 features and the specific 3d orbitals of Mn^{3+}, Mn^{4+}, and Co^{3+} in octahedral sites, the quantitative molecular orbital diagram is generated in Figure 5 by considering crystal field and spin states. For high-spin Mn^{4+} ($3d^3$, t_{2g}^3) in Li_2MnO_3, only two peaks were observed from the transitions to t_{2g}-down/e_g-up (~530.2 eV) and e_g-down (~532.4 eV) states. For high-spin Mn^{3+} ($3d^4$, $t_{2g}^3e_g^1$) in Mn_2O_3, triple-peak structure was observed from the transitions to e_g-up (~530.4 eV), t_{2g}-down (~531.4 eV), and e_g-down (533.2 eV) states [36,40]. For low-spin Co^{3+} ($3d^6$, t_{2g}^6) in $LiCoO_2$, two-peak structure was observed from the transitions to e_g-up (~531.2 eV) and e_g-down (532.6 eV) states. Considering the energy position of the Gaussian peaks and the specific 3d orbital of the references in Figure 5, we assign the partial covalency of specific orbital to the four Gaussian peaks of pre-edge portion. For P2 feature, two sources of partial covalency are Co^{3+} 3d e_g-up and Mn^{3+} 3d t_{2g}-down orbitals. However, the Mn^{3+} 3d t_{2g}-down orbital contributes little to this energy position from the O K-edge in Figure 4a. Therefore, the main source of partial covalency is Co^{3+} 3d e_g-up orbital for P2. Similarly, since Co^{3+} 3d e_g-down orbital contributes little to the P3 energy position, the main source of partial covalency is the Mn^{4+} 3d e_g-down orbital for P3. Finally, the relationship between the four Gaussian features and main partial covalency is concluded in Table 1.

Nanomaterials **2019**, *9*, 577

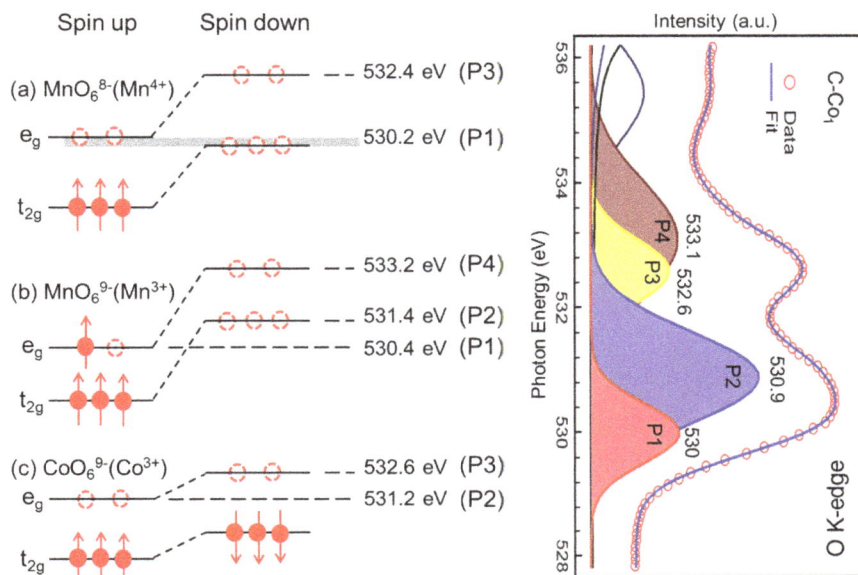

Figure 5. The correlation between four Gaussian peaks of pre-edge portion and partial covalency of specific 3d orbitals. (**Left**) Experimental molecular orbital diagrams based on the XAS spectra. The three clusters, (**a**) MnO_6^{8-}, (**b**) MnO_6^{9-}, and (**c**) CoO_6^{9-} correspond to Mn^{4+}, Mn^{3+}, and Co^{3+} in octahedral structure, respectively. (**Right**) O–K pre-edge XAS signals with peak decomposition of C–Co_1.

Table 1. The correspondence between the Gaussian peak and partial covalency.

Gaussian Peak (Energy Position)	Main Partial Covalency
P1 (~530.0 eV)	Mn^{4+} 3d t_{2g}-down and e_g-up orbital, Mn^{3+} 3d e_g-up orbital
P2 (~530.9 eV)	Co^{3+} 3d e_g-up orbital
P3 (~532.6 eV)	Mn^{4+} 3d e_g-down orbital
P4 (~533.1e V)	Mn^{3+} 3d e_g-down orbital

After distinguishing the partial covalency, its relationship with the ORR activity is analyzed. Since the valence of the metals has little change, the area of the Gaussian feature is mainly determined by the metal ratio and the covalent strength. We normalize the Gaussian feature area to the metal ratio to represent the covalent strength. The relationship between the strength of partial covalency with the half-wave potential $E_{1/2}$ for the cubic and tetragonal spinel oxides shows in Figure 6a,b, respectively. For both cubic and tetragonal phase, increasing the covalent strength of P1 and P2 positively affects ORR activity, while increasing the covalent strength of P3 and P4 negatively affects ORR activity in tetragonal and cubic phase, respectively. This illustrates that the partial covalency of P1 and P2 can boost the ORR catalytic activities, while the partial covalency of P3 and P4 may demote the ORR catalytic activities. As displayed in Table 1, the main origin of the partial covalency of P1 is Mn^{4+} 3d t_{2g}-down/e_g-up and Mn^{3+} 3d e_g-up orbitals, and P2 is Co^{3+} 3d e_g-up orbital. That is to say, the enhanced ORR catalytic activities can be attributed to the raised partial covalent strength of O 2p orbital with Mn^{4+} 3d t_{2g}-down/e_g-up, Mn^{3+} 3d e_g-up, and Co^{3+} 3d e_g-up orbitals in this spinel system.

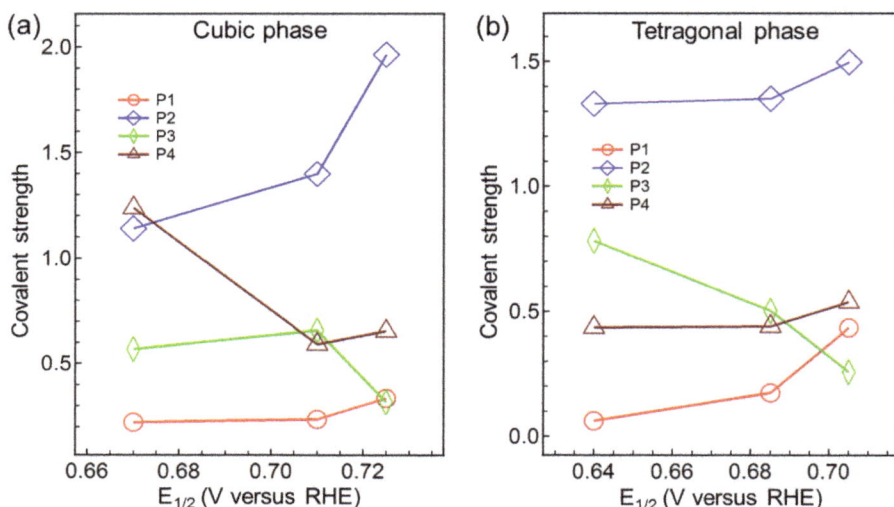

Figure 6. The relationship between the partial M 3d-O 2p covalency strength and ORR catalytic activities. The covalency strength of P1, P2, P3, and P4 versus the half-wave potential $E_{1/2}$ in the cubic (**a**) and tetragonal (**b**) phase.

Previous researches have demonstrated that the stronger covalent strength of M 3d-O 2p, the more ORR electrocatalytic activity of the oxide can be promoted [19,24], where the covalent strength of M–O are thought to have a critical impact on the rate of desorption and adsorption steps. Our study indicates that not all the covalency of metal 3d orbital and O 2p orbital have a positive push effect on ORR activity. Specifically, the partial covalency of O 2p orbital with Mn^{4+} 3d t_{2g}-down/e_g-up, Mn^{3+} 3d e_g-up, and Co^{3+} 3d e_g-up orbitals is benefit for ORR activity. It is consistent with the common sense that the e_g orbitals of the metal, rather than the t_{2g} orbitals, interact more easily with the oxygen orbital to produce an adsorbed intermediate with end-on absorption mode [44,45]. In addition, the single electron in Mn^{3+} 3d e_g-up orbital is also important for the improving catalytic activity. As discussed in previous reports, the existence of a single e_g electron in the TM ion is able to form a covalent interaction with the adsorbate [46] and the proper number of e_g orbital electron filling can well modulate the balance of intermediates displacement in rate-limiting reactions during traditional four-electron ORR proceeds (Figure S2 in Supplementary Materials) [19–22]. On account of the above discussion, variations in the ORR performance between $Co_xMn_{3-x}O_4$ (x = 1, 1.5, 2) can be modified by the partial covalency of TM 3d−O 2p by tuning the interaction between spinel catalyst and the molecular oxygen. This may open up a new way to find high-efficiency ORR catalysts by introducing other metals to tune the partial covalent states of M 3d-O 2p.

4. Conclusions

In summary, we have studied a substitution strategy to tune the electronic structure of $Co_xMn_{3-x}O_4$ in terms of TM 3d-O 2p covalency, especially the partial covalency between Mn^{4+} 3d t_{2g}-down/e_g-up, Mn^{3+} 3d e_g-up, and Co^{3+} 3d e_g-up orbitals with O 2p orbital, for enhancing the ORR activity. The ORR performance increases with the increase of Mn content in both cubic and tetragonal phases. The electronic structure of metal (cobalt and manganese) has very little changes for the substitution of Co by Mn, however, the electronic structure of O can be significantly regulated by studying sXAS. In other words, the factor determining the ORR activity are related to the M 3d-O 2p covalency. More important, the partial covalency between Mn^{4+} 3d t_{2g}-down/e_g-up, Mn^{3+} 3d e_g-up, and Co^{3+} 3d e_g-up orbitals with O 2p orbital plays an significant role on the ORR catalytic activities by further analysis of O K-edge, such as integration and peak deconvolution. Our findings demonstrate the key role of partial covalency between specific metal 3d orbitals and O 2p orbital in enhancing the ORR activity, which can provide a new avenue to designing high efficiency catalyst materials for clean energy conversion and storage devices.

Supplementary Materials: The following are available online at http://www.mdpi.com/2079-4991/9/4/577/s1. Figure S1: The O K-edge sXAS signals of the spinel $Co_xMn_{3-x}O_4$ oxides. After subtracting an arctangent background, eight Gaussian functions were employed to fit the sXAS spectrum. Figure S2: Proposed traditional four-electron ORR mechanism on spinel oxide catalysts. The ORR proceeds via four steps: 1, surface oxygen gas adsorption; 2, surface peroxide formation; 3, surface oxide formation; 4, surface hydroxide regeneration. M is a transition-metal cation in octahedral sites. Table S1: The full width at half maximum (FWHM) of the Gaussian functions and the arctangent background (ATAN function) for the O K-edge sXAS spectra peak deconvolution of the cubic and tetragonal spinel oxides (unit: eV). Table S2: The energy position of the Gaussian functions and ATAN function for the O K-edge sXAS spectra peak deconvolution of the cubic and tetragonal spinel oxides (unit: eV).

Author Contributions: Conceptualization, X.L. (Xiaosong Liu); Data curation, X.L. (Xinghui Long), P.Y., N.Z., X.F., G.R., S.Z., and J.F.; Formal analysis, P.Y., N.Z., X.F., and F.C.; Funding acquisition, X.L. (Xiaosong Liu); Investigation, X.L. (Xinghui Long); Methodology, X.L. (Xinghui Long), P.Y., N.Z., and X.F.; Project administration, X.L. (Xiaosong Liu); Resources, C.L. and F.C.; Software, X.L. (Xinghui Long); Supervision, X.L. (Xiaosong Liu); Writing—original draft, X.L. (Xinghui Long), P.Y. and N.Z.; Writing—review & editing, X.L. (Xiaosong Liu).

Funding: This research was funded by the National Natural Science Foundation of China, grant number 21473235, U1632269, 11227902, and 21503263.

Acknowledgments: The authors thank Jun Chen at Nankai University for fruitful discussion on cubic and tetragonal $Co_xMn_{3-x}O_4$ (x = 1, 1.5, 2) synthesis and ORR electrocatalysis experiments. The authors also thank Jenn-Min Lee and Jin-Ming Chen for helpful discussions about sXAS experiment and analysis and synchrotron Taiwan Light Source (TLS) for providing beamtime at the 20A1 beamline.

Conflicts of Interest: The authors declare no conflict of interest.

References

1. Gray, H.B. Powering the Planet with Solar Fuel. *Nat. Chem.* **2009**, *1*, 7. [CrossRef] [PubMed]
2. Dau, H.; Limberg, C.; Reier, T.; Risch, M.; Roggan, S.; Strasser, P. The Mechanism of Water Oxidation: From Electrolysis via Homogeneous to Biological Catalysis. *ChemCatChem* **2010**, *2*, 724–761. [CrossRef]
3. Wang, Z.L.; Xu, D.; Xu, J.J.; Zhang, X.B. Oxygen electrocatalysts in metal-air batteries: From aqueous to nonaqueous electrolytes. *Chem. Soc. Rev.* **2014**, *43*, 7746–7786. [CrossRef] [PubMed]
4. Service, R.F. Hydrogen Cars: Fad or the Future? *Science* **2009**, *324*, 1257–1259. [CrossRef] [PubMed]
5. Chen, C.; Kang, Y.; Huo, Z.; Zhu, Z.; Huang, W.; Xin, H.L.; Snyder, J.D.; Li, D.; Herron, J.A.; Mavrikakis, M.; et al. Highly crystalline multimetallic nanoframes with three-dimensional electrocatalytic surfaces. *Science* **2014**, *343*, 1339–1343. [CrossRef]
6. Lee, Y.; Suntivich, J.; May, K.J.; Perry, E.E.; Shao-Horn, Y. Synthesis and Activities of Rutile IrO_2 and RuO_2 Nanoparticles for Oxygen Evolution in Acid and Alkaline Solutions. *J. Phys. Chem. Lett.* **2012**, *3*, 399–404. [CrossRef]
7. Luo, Z.; Marti-Sanchez, S.; Nafria, R.; Joshua, G.; de la Mata, M.; Guardia, P.; Flox, C.; Martinez-Boubeta, C.; Simeonidis, K.; Llorca, J.; Morante, J.R.; Arbiol, J.; Ibanez, M.; Cabot, A. $Fe_3O_4@NiFe_xO_y$ Nanoparticles with Enhanced Electrocatalytic Properties for Oxygen Evolution in Carbonate Electrolyte. *ACS Appl. Mater. Interfaces* **2016**, *8*, 29461–29469. [CrossRef]
8. Davodi, F.; Tavakkoli, M.; Lahtinen, J.; Kallio, T. Straightforward synthesis of nitrogen-doped carbon nanotubes as highly active bifunctional electrocatalysts for full water splitting. *J. Catal.* **2017**, *353*, 19–27. [CrossRef]
9. Davodi, F.; Muhlhausen, E.; Tavakkoli, M.; Sainio, J.; Jiang, H.; Gokce, B.; Marzun, G.; Kallio, T. Catalyst Support Effect on the Activity and Durability of Magnetic Nanoparticles: Toward Design of Advanced Electrocatalyst for Full Water Splitting. *ACS Appl. Mater. Interfaces* **2018**, *10*, 31300–31311. [CrossRef]
10. Rios, E.; Gautier, J.-L.; Poillerat, G.; Chartier, P. Mixed Valency Spinel Oxides of Transition Metals and Electrocatalysis: Case of the $Mn_xCo_{3-x}O_4$ System. *Electrochim. Acta* **1998**, *44*, 1491–1497. [CrossRef]

11. Ponce, J.; Rehspringer, J.-L.; Poillerat, G.; Gautier, J.L. Electrochemical study of nickel–aluminium–manganese spinel $Ni_xAl_{1-x}Mn_2O_4$. Electrocatalytical properties for the oxygen evolution reaction and oxygen reduction reaction in alkaline media. *Electrochim. Acta* **2001**, *46*, 3373–3380. [CrossRef]

12. Cui, B.; Lin, H.; Li, J.-B.; Li, X.; Yang, J.; Tao, J. Core-Ring Structured $NiCo_2O_4$ Nanoplatelets: Synthesis, Characterization, and Electrocatalytic Applications. *Adv. Funct. Mater.* **2008**, *18*, 1440–1447. [CrossRef]

13. Li, Y.; Hasin, P.; Wu, Y. $Ni_xCo_{3-x}O_4$ nanowire arrays for electrocatalytic oxygen evolution. *Adv. Mater.* **2010**, *22*, 1926–1929. [CrossRef]

14. Wang, H.; Yang, Y.; Liang, Y.; Zheng, G.; Li, Y.; Cui, Y.; Dai, H. Rechargeable $Li–O_2$ batteries with a covalently coupled $MnCo_2O_4$–graphene hybrid as an oxygen cathode catalyst. *Energy Environ. Sci.* **2012**, *5*, 7931. [CrossRef]

15. Zhang, L.; Zhang, S.; Zhang, K.; Xu, G.; He, X.; Dong, S.; Liu, Z.; Huang, C.; Gu, L.; Cui, G. Mesoporous $NiCo_2O_4$ nanoflakes as electrocatalysts for rechargeable $Li-O_2$ batteries. *Chem. Commun.* **2013**, *49*, 3540–3542. [CrossRef] [PubMed]

16. Kim, T.W.; Woo, M.A.; Regis, M.; Choi, K.-S. Electrochemical Synthesis of Spinel Type $ZnCo_2O_4$ Electrodes for Use as Oxygen Evolution Reaction Catalysts. *J. Phys. Chem. Lett.* **2014**, *5*, 2370–2374. [CrossRef] [PubMed]

17. Rios, E.; Chartier, P.; Gautier, J.-L. Oxygen Evolution Electrocatalysis in Alkaline Medium at thin $Mn_xCo_{3-x}O_4$ $(0 < x < 1)$ Spinel Films on Glass/SnO_2: F Prepared by Spray Pyrolysis. *Solid State Sci.* **1999**, *1*, 267–277.

18. Restovic, A.; Rı'os, E.; Barbato, S.; Ortiz, J.; Gautier, J.L. Oxygen Reduction in Alkaline Medium at Thin $Mn_xCo_{3-x}O_4$ $(0<x<1)$ Spinel Films Prepared by Spray Pyrolysis. Effect of Oxide Cation Composition on the Reaction Kinetics. *J. Electroanal. Chem.* **2002**, *522*, 141–151.

19. Suntivich, J.; Gasteiger, H.A.; Yabuuchi, N.; Nakanishi, H.; Goodenough, J.B.; Shao-Horn, Y. Design principles for oxygen-reduction activity on perovskite oxide catalysts for fuel cells and metal-air batteries. *Nat. Chem.* **2011**, *3*, 546–550. [CrossRef]

20. Suntivich, J.; May, K.J.; Gasteiger, H.A.; Goodenough, J.B.; Shao-Horn, Y. A perovskite oxide optimized for oxygen evolution catalysis from molecular orbital principles. *Science* **2011**, *334*, 1383–1385. [CrossRef] [PubMed]

21. Wei, C.; Feng, Z.; Scherer, G.G.; Barber, J.; Shao-Horn, Y.; Xu, Z.J. Cations in Octahedral Sites: A Descriptor for Oxygen Electrocatalysis on Transition-Metal Spinels. *Adv. Mater.* **2017**, *29*, 1606800. [CrossRef]

22. Zhou, Y.; Sun, S.; Xi, S.; Duan, Y.; Sritharan, T.; Du, Y.; Xu, Z.J. Superexchange Effects on Oxygen Reduction Activity of Edge-Sharing $[Co_xMn_{1-x}O_6]$ Octahedra in Spinel Oxide. *Adv. Mater.* **2018**, *30*, 1705407. [CrossRef] [PubMed]

23. Mueller, D.N.; Machala, M.L.; Bluhm, H.; Chueh, W.C. Redox activity of surface oxygen anions in oxygen-deficient perovskite oxides during electrochemical reactions. *Nat. Commun.* **2015**, *6*, 6097. [CrossRef] [PubMed]

24. Suntivich, J.; Hong, W.T.; Lee, Y.-L.; Rondinelli, J.M.; Yang, W.; Goodenough, J.B.; Dabrowski, B.; Freeland, J.W.; Shao-Horn, Y. Estimating Hybridization of Transition Metal and Oxygen States in Perovskites from O K-edge X-ray Absorption Spectroscopy. *J. Phys. Chem. C* **2014**, *118*, 1856–1863. [CrossRef]

25. Cheng, F.; Shen, J.; Peng, B.; Pan, Y.; Tao, Z.; Chen, J. Rapid room-temperature synthesis of nanocrystalline spinels as oxygen reduction and evolution electrocatalysts. *Nat. Chem.* **2011**, *3*, 79–84. [CrossRef] [PubMed]

26. Li, C.; Han, X.; Cheng, F.; Hu, Y.; Chen, C.; Chen, J. Phase and composition controllable synthesis of cobalt manganese spinel nanoparticles towards efficient oxygen electrocatalysis. *Nat. Commun.* **2015**, *6*, 7345. [CrossRef] [PubMed]

27. Chung, S.C.; Chen, C.I.; Tseng, P.C.; Lin, H.F.; Dann, T.E.; Song, Y.F.; Huang, L.R.; Chen, C.C.; Chuang, J.M.; Tsang, K.L.; et al. Soft X-ray spectroscopy beamline 6 m high energy spherical grating monochromator at SRRC: Optical design and first performance tests. *Rev. Sci. Instrum.* **1995**, *66*, 1655–1657. [CrossRef]

28. Morales, F.; de Groot, F.M.F.; Glatzel, P.; Kleimenov, E.; Bluhm, H.; Havecker, M.; Knop-Gericke, A.; Weckhuysen, B.M. In Situ X-ray Absorption of $Co/Mn/TiO_2$ Catalysts for Fischer-Tropsch Synthesis. *J. Phys. Chem. B* **2004**, *108*, 16201–16207. [CrossRef]

29. Ghiasi, M.; Delgado-Jaime, M.U.; Malekzadeh, A.; Wang, R.-P.; Miedema, P.S.; Beye, M.; de Groot, F.M.F. Mn and Co Charge and Spin Evolutions in $LaMn_{1-x}Co_xO_3$ Nanoparticles. *J. Phys. Chem. C* **2016**, *120*, 8167–8174. [CrossRef]

Nanomaterials **2019**, *9*, 577

30. Lin, F.; Markus, I.M.; Nordlund, D.; Weng, T.C.; Asta, M.D.; Xin, H.L.; Doeff, M.M. Surface reconstruction and chemical evolution of stoichiometric layered cathode materials for lithium-ion batteries. *Nat. Commun.* **2014**, *5*, 3529. [CrossRef] [PubMed]

31. Chang, C.F.; Hu, Z.; Wu, H.; Burnus, T.; Hollmann, N.; Benomar, M.; Lorenz, T.; Tanaka, A.; Lin, H.J.; Hsieh, H.H.; et al. Spin Blockade, Orbital Occupation, and Charge Ordering in $La_{1.5}Sr_{0.5}CoO_4$. *Phys. Rev. Lett.* **2009**, *102*, 116401. [CrossRef] [PubMed]

32. Suchow, L. A Detailed, Simple Crystal Field Consideration of the Normal Spinel Structure of Co_3O_4. *J. Chem. Educ.* **1976**, *53*, 560. [CrossRef]

33. Kim, K.J.; Park, Y.R. Optical investigation of charge-transfer transitions in spinel Co_3O_4. *Solid State Commun.* **2003**, *127*, 25–28. [CrossRef]

34. Liang, Y.; Wang, H.; Zhou, J.; Li, Y.; Wang, J.; Regier, T.; Dai, H. Covalent hybrid of spinel manganese-cobalt oxide and graphene as advanced oxygen reduction electrocatalysts. *J. Am. Chem. Soc.* **2012**, *134*, 3517–3523. [CrossRef] [PubMed]

35. Qiao, R.; Wang, Y.; Olalde-Velasco, P.; Li, H.; Hu, Y.-S.; Yang, W. Direct evidence of gradient Mn(II) evolution at charged states in $LiNi_{0.5}Mn_{1.5}O_4$ electrodes with capacity fading. *J. Power Sources* **2015**, *273*, 1120–1126. [CrossRef]

36. Qiao, R.; Chin, T.; Harris, S.J.; Yan, S.; Yang, W. Spectroscopic fingerprints of valence and spin states in manganese oxides and fluorides. *Curr. Appl. Phys.* **2013**, *13*, 544–548. [CrossRef]

37. Cramer, S.P.; DeGroot, F.M.F.; Ma, Y.; Chen, C.T.; Sette, F.; Kipke, C.A.; Eichhorn, D.M.; Chan, M.K.; Armstrong, W.H. Ligand Field Strengths and Oxidation States From Manganese L-Edge Spectroscopy. *J. Am. Chem. Soc.* **1991**, *113*, 7937–7940. [CrossRef]

38. Thole, B.T.; Vanderlaan, G. Branching Ratio in X-ray Absorption Spectroscopy. *Phys. Rev. B* **1988**, *38*, 3158–3171. [CrossRef]

39. Ralston, C.Y.; Wang, H.; Ragsdale, S.W.; Kumar, M.; Spangler, N.J.; Ludden, P.W.; Gu, W.; Jones, R.M.; Patil, D.S.; Cramer, S.P. Characterization of Heterogeneous Nickel Sites in CO Dehydrogenases from Clostridium thermoaceticum and Rhodospirillum rubrum by Nickel L-Edge X-ray Spectroscopy. *J. Am. Chem. Soc.* **2000**, *122*, 10553–10560. [CrossRef]

40. Kurata, H.; Colliex, C. Electron-energy-loss core-edge structures in manganese oxides. *Phys. Rev. B* **1993**, *48*, 2102–2108. [CrossRef]

41. De Groot, F.M.F.; Grioni, M.; Fuggle, J.C.; Ghijsen, J.; Sawatzky, G.A.; Petersen, H. Oxygen 1s X-ray-Absorption Edges of Transition-Metal Oxides. *Phys. Rev. B* **1989**, *40*, 5715–5723. [CrossRef]

42. Minasian, S.G.; Keith, J.M.; Batista, E.R.; Boland, K.S.; Bradley, J.A.; Daly, S.R.; Kozimor, S.A.; Lukens, W.W.; Martin, R.L.; Nordlund, D.; et al. Covalency in metal-oxygen multiple bonds evaluated using oxygen K-edge spectroscopy and electronic structure theory. *J. Am. Chem. Soc.* **2013**, *135*, 1864–1871. [CrossRef] [PubMed]

43. Ye, Y.; Thorne, J.E.; Wu, C.H.; Liu, Y.S.; Du, C.; Jang, J.W.; Liu, E.; Wang, D.; Guo, J. Strong O 2p-Fe 3d Hybridization Observed in Solution-Grown Hematite Films by Soft X-ray Spectroscopies. *J. Phys. Chem. B* **2018**, *122*, 927–932. [CrossRef] [PubMed]

44. Matsumoto, Y.; Yoneyama, H.; Tamura, H. Influence of the Nature of the Conduction Band of Transition Metal Oxides on Catalytic Activity for Oxygen Reduction. *J. Electroanal. Chem.* **1977**, *83*, 237–243. [CrossRef]

45. Yeager, E. Dioxygen Electrocatalysis: Mechanisms in Relation to Catalyst Structure. *J. Mol. Catal.* **1986**, *38*, 5–25. [CrossRef]

46. Ahmad, E.A.; Tileli, V.; Kramer, D.; Mallia, G.; Stoerzinger, K.A.; Shao-Horn, Y.; Kucernak, A.R.; Harrison, N.M. Optimizing Oxygen Reduction Catalyst Morphologies from First Principles. *J. Phys. Chem. C* **2015**, *119*, 16804–16810. [CrossRef]

nanomaterials

MDPI

Article

Elemental Distribution and Structural Characterization of GaN/InGaN Core-Shell Single Nanowires by Hard X-ray Synchrotron Nanoprobes

Eleonora Secco [1]**, Heruy Taddese Mengistu** [1]**, Jaime Segura-Ruíz** [2]**, Gema Martínez-Criado** [2,3]**, Alberto García-Cristóbal** [1]**, Andrés Cantarero** [4]**, Bartosz Foltynski** [5]**, Hannes Behmenburg** [5]**, Christoph Giesen** [5]**, Michael Heuken** [5] **and Núria Garro** [1,*]

[1]   Institut de Ciència dels Materials (ICMUV), Universitat de València, 46980 Paterna (València), Spain; eleonora.secco@uv.es (E.S.); Heruy.Mengistu@uv.es (H.T.M.); alberto.garcia@uv.es (A.G.-C.)
[2]   ESRF—The European Synchrotron, 71 avenue des Martyrs, 38043 Grenoble, France; jaime.segura@esrf.fr (J.S.-R.); gema.martinez.criado@csic.es (G.M.-C.)
[3]   Instituto de Ciencia de Materiales de Madrid (ICMM), Consejo Superior de Investigaciones Científicas (CSIC), Sor Juana Inés de la Cruz 3, 28049 Madrid, Spain
[4]   Institut de Ciència Maolecular (ICMOL), Universitat de València, 46980 Paterna (València), Spain; andres.cantarero@uv.es
[5]   AIXTRON SE, Dornkaulstrasse 2, 52134 Herzogenrath, Germany; B.Foltynkski@aixtron.com (B.F.); H.Behmenburg@aixtron.com (H.B.); C.Giesen@aixtron.com (C.G.); M.Heuken@aixtron.com (M.H.)
*   Correspondence: nuria.garro@uv.es; Tel.: +34-9635-436-01

Received: 28 March 2019; Accepted: 24 April 2019; Published: 3 May 2019

Abstract: Improvements in the spatial resolution of synchrotron-based X-ray probes have reached the nano-scale and they, nowadays, constitute a powerful platform for the study of semiconductor nanostructures and nanodevices that provides high sensitivity without destroying the material. Three complementary hard X-ray synchrotron techniques at the nanoscale have been applied to the study of individual nanowires (NWs) containing non-polar GaN/InGaN multi-quantum-wells. The trace elemental sensitivity of X-ray fluorescence allows one to determine the In concentration of the quantum wells and their inhomogeneities along the NW. It is also possible to rule out any contamination from the gold nanoparticle catalyst employed during the NW growth. X-ray diffraction and X-ray absorption near edge-structure probe long- and short-range order, respectively, and lead us to the conclusion that while the GaN core and barriers are fully relaxed, there is an induced strain in InGaN layers corresponding to a perfect lattice matching with the GaN core. The photoluminescence spectrum of non-polar InGaN quntum wells is affected by strain and the inhomogeneous alloy distribution but still exhibits a reasonable 20% relative internal quantum efficiency.

Keywords: semiconductor nanowires; synchrotron probes; nano-scale resolution

1. Introduction

Nanowire based GaN/InGaN heterostructures, such as quantum wells (QWs), are acknowledged as optimum support for highly efficient optoelectronic devices [1]. This is mostly based on two facts: first of all, the wide spectral range potentially covered by InGaN QWs, which suits perfectly the requirements of lighting and photovoltaic solar cells [2] and, on the other hand, the benefits of NW morphology in overcoming the main difficulties affecting planar geometry [3]. The latter include the drastic reduction of the compositional inhomogeneities and segregation affecting indium [4], the more effective strain release [5], and the possibility to grow heterostructures on the NW lateral surfaces (corresponding to non-polar *m*-planes), which expands the optically active area and increases its quantum efficiency [6]. The significant efforts dedicated to obtain GaN/InGaN core-shell NWs with high structural and optical quality have nowadays fructified with the demonstration of several NW-based devices, such as light emitting diodes [7,8], laser diodes [9], and photodiodes [10].

Since device performance relies on the composition, homogeneity, and quality of the heterostructures, a careful and complete characterization with focus on individual GaN/InGaN NWs is a crucial milestone in this field. Techniques that enable the study of single NWs with spatial resolutions ranging from the micro- to the nano-scale are required. Moreover, in nanoscale functional materials, it becomes very important to correlate structural and optical properties by applying different techniques to the same nano-object. Therefore, the use of non-destructive and contact-less techniques is mandatory. For a long time, the most employed technique for the study of the core-shell NWs had been high resolution scanning transmission electron microscopy (HR-STEM) [11–13]. The analysis of indium distribution and concentration in the QWs could be carried out by energy-dispersive X-ray spectroscopy (EDS) [14,15] with high spatial resolution but low detection limits at NW level. These measurements, however, tend to be time-costly; difficult; and, most importantly, imply the destruction of a part of the sample.

The nanometer scale spatial resolution achieved in recent years in third-generation synchrotrons enables the study of single NWs with complementary techniques, such as X-ray fluorescence (XRF), X-ray diffraction (XRD), and X-ray absorption near edge structure (XANES). These are key tools for the determination of the chemical composition and structural properties of materials in a non-destructive manner with high sensitivity and selectivity and, nowadays, with sufficient spatial resolution. Furthermore, these allow for the scanning the X-ray beam and performing maps of single NWs. These clear benefits have expanded the use of synchrotron X-ray nanoprobe to the investigation of single InGaN NWs [16–19], even to those including non-polar InGaN/GaN QWs [20–22].

This work aims to demonstrate the benefits of synchrotron-based X-ray spectroscopies for the study of GaN/InGaN core-shell NWs grown by catalyst-assisted metal-organic chemical vapor deposition (MOCVD) containing non-polar QWs on their lateral surfaces. These systems typically present inhomogeneities in their elemental composition, defect concentration, and strain fields at the nanoscale, and their characterization by conventional techniques has proved to be complex and costly. A hard X-ray synchrotron nanoprobe provides the multi-technique platform required for this kind of studies. The elemental composition and their distribution were characterized by XRF. Strain fields and crystal quality were probed by XRD. Local order effects and elemental segregation could be analysed by XANES spectroscopy. The information gathered by the combination of these complementary techniques can be then used to interpret the photoluminescence (PL) spectra of individual NWs. Finally, numerical simulations reinforce the consistency of the experimental results.

2. Materials and Methods

Core-shell GaN/InGaN multiple-QWs (MQWs) on GaN NWs were grown by MOCVD in an AIXTRON 3 × 2" close-coupled showerhead reactor (AIXTRON SE, Herzogenrath, Germany). Details about the growth procedure are included in the supplementary information. A scanning electron microscopy (SEM) bird's eye view micrograph of the sample is shown in Figure 1a, while a high resolution transmission electron microscopy (HR-TEM) image of a single dispersed NW is presented in Figure 1b. The NWs have diameters and heights ranging from 150 to 250 nm and 1.5 to 3 μm, respectively, and their longitudinal axis coincides with the *c*-axis of the wurtzite structure. Three GaN/InGaN MQWs form on the lateral non-polar *m*-planes of the NWs, whose expected structure is depicted in Figure 1c. In order to check whether the MQWs formed correctly during the growth process, energy-dispersive X-ray spectroscopy (EDS) measurements were performed along axial and radial directions on individual NWs. Figure 1d represents the integrated intensity of the In L_α and Ga K_α fluorescence lines recorded at each point of the scan along the radial direction of a NW. The core-shell MQWs and the barriers appear as regular oscillations in the In and Ga fluorescence signal. Linear EDS scans were also performed along the axial direction of the NW (see supplementary Figure S1), which indicates that no polar QWs formed on the top NW surface. By taking advantage of the 4 nm spatial resolution of EDS, we estimate the widths of the wells and the barriers by fitting the spectra of the lateral scans to the structure depicted in Figure 1c. This results in the following values: $R = 93 \pm 1.3$ nm, $t = 5.7 \pm 3.2$ nm, and $s = 4.9 \pm 2.3$ nm. With these dimensions, the core represents 57% of the NW volume, while the shell takes the remaining 43%. On the other hand, quantifying the In concentration of the non-polar QWs is dissuasive due to the low signal to noise ratio of the EDS signal.

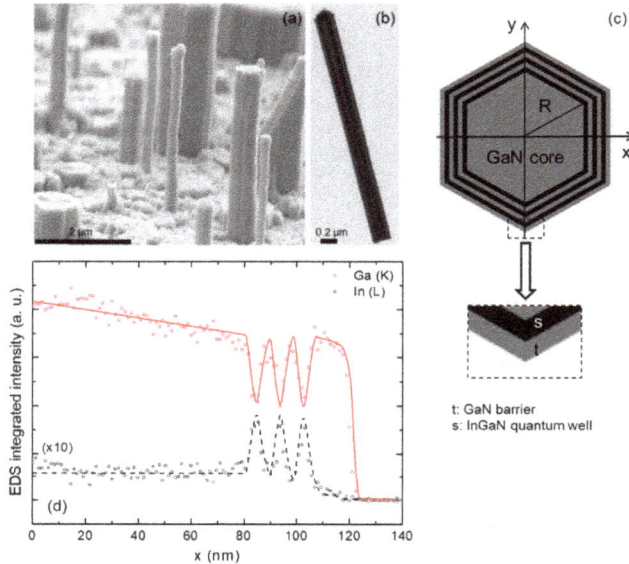

Figure 1. (**a**) SEM view of the as-grown nanowires (NWs) of the sample characterized in this study. (**b**) high resolution transmission electron microscopy (HR-TEM) of the single dispersed NW scanned by energy-dispersive X-ray spectroscopy (EDS). (**c**) Scheme of the cross-section of the core-shell GaN/InGaN NW and the magnification of its upper corner where *t* and *s* are the thickness of the InGaN quantum wells (QWs) and the GaN barriers, respectively. (**d**) Radial scans of the integrated intensities of the In L_α and Ga K_α peaks of the EDS spectrum taken at the top of the NW. The lines are the corresponding fitting curves.

Individual NWs were dispersed on 200 nm-thick SiN windows for XRF, XRD, and XANES measurements at the nanoimaging station ID22NI (currently ID16B) of the European Synchrotron Radiation Facility (ESRF, Grenoble, France) with a multitechnique setup similar to the one described by J. Segura-Ruiz el al. [19]. The pink X-ray beam ($\Delta E/E \sim 10^{-2}$) was focused providing a spot of 120 $\times$ 97 nm^2 at an energy of 29.6 keV. The X-ray beam impinged nearly perpendicular to the sample surface and the XRF signal was detected at 15° using a single element energy dispersive silicon drift detector. XRF spectra were analysed using the program PYMCA [23]. XRD measurements were carried out with a monochromatic beam (($\Delta E/E \sim 10^{-4}$), with a spot size of 154 $\times$ 136 nm^2 and an energy of 28.029 keV. The XRD signal was measured using a large field of view (94 $\times$ 94 mm^2) fast readout low noise (FReLoN) charge coupled device (CCD) detector F-K4320T equipped with 3.3/1 demagnifying fibre optics taper (ESRF, Grenoble, France). The XRD data were analysed using both the Fit2D package [24] and the PYMCA program. Fit2D allows one to calibrate standard diffraction patterns, using the experimental parameters derived from the measurement of an Al$_2$O$_3$ reference sample. XANES spectra were recorded in XRF mode with a step size of 1 eV (equaling the resolution of the Si (111) double crystal monochromator) and an integration time of 1 s/point. The data were normalised and analysed with the IFFEFIT package [25].

Raman scattering and photoluminescence (PL) measurements were obtained using an optical microscope coupled to a T64000 triple spectrometer (Horiba Jobin-Yvon®, France) and a nitrogen-cooled CCD. The spectral resolution of the whole system was of 1 cm^{-1} at 500 nm. Microscope objectives of 100 and 40 magnification provided laser spot diameters of 1 and 4 µm for visible and UV excitation, respectively. The Raman spectra were collected in back-scattering configuration at room temperature with the 514 nm line of an Ar laser. PL measurements were performed at controlled temperature (7–300 K) with a 325 nm He-Cd laser.

3. Results

3.1. Composition and Structural Properties of Individual NWs by Raman Scattering

Raman scattering can provide a first insight into In concentration and structural quality of the NWs. Micro-Raman measurements, with a spatial resolution of 1 µm, probe individual GaN/InGaN core-shell NWs that have been previously dispersed on an inert substrate. A representative Raman spectrum is shown in Figure 2. The observed Raman peaks have been fitted by Gaussians; their values match those of wurtzite strain-free GaN and should come mostly from the NW core. Due to the oblique facets of the NW, Raman selection rules are relaxed and four modes attributed to wurtzite GaN are observed: A_1 (TO), E_1 (TO), E_{2h}, and E_1 (LO). The FWHMs of the E_{2h} mode measured on several NWs are approximately 3.5 cm^{-1}, indicating the good structural quality of the NWs. An additional peak centered around 701 cm^{-1} could match the frequency of the optical mode of the InGaN QWs. We cannot disregard, however, that this mode could correspond to a surface optical mode (SO), which typically appears between the TO and LO frequencies and is significantly intense in NWs due to the large surface-to-volume ratio [26]. Thus, no definitive conclusions are drawn regarding the In concentration and strain of the QWs from Raman investigations.

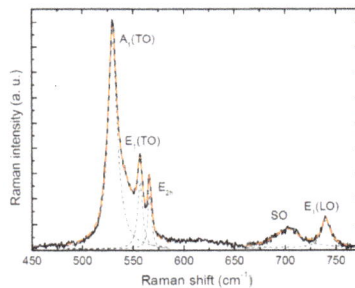

Figure 2. Representative unpolarized Raman spectrum of a single NW with its best fit (dashed lines).

3.2. Elemental Distribution and In Concentration along Individual NWs by X-ray Fluorescence

The composition of several GaN/InGaN core-shell NWs dispersed on a SiN thin window has been investigated recording XRF maps of the area enclosing the NW and its surrounding region. Figure 3a shows the XRF spectra measured at different regions of an individual NW: one at the top and at the middle of the NW, and at a location corresponding to the SiN window. Each spectra is obtained by averaging over 10 pixels (each pixel is 25 × 25 nm^2) in the different regions of the map. The most intense peaks in the spectra are those corresponding to In, Ga, Au, and Ag. Outside the NW, Ga and In XRF intensities have vanished almost completely and an additional peak attributed to Pb is observed. The presence of weak Ga spectral peaks outside the NW can be justified by the long decaying tails (vertical and horizontal) of the non-circular X-ray beam and the low detection limits of this technique. Having a Pb signal is not surprising either, and it comes from the shielding system around the sample. Au and Ag were detected only at the top of the NWs and are attributed to the catalyst used for the NW growth. Ag, which can be present in very low levels in Au, has an estimated concentration of around 0.02% as calculated by PYMCA. This figure confirms the trace chemical sensitivity of synchrotron nano-XRF.

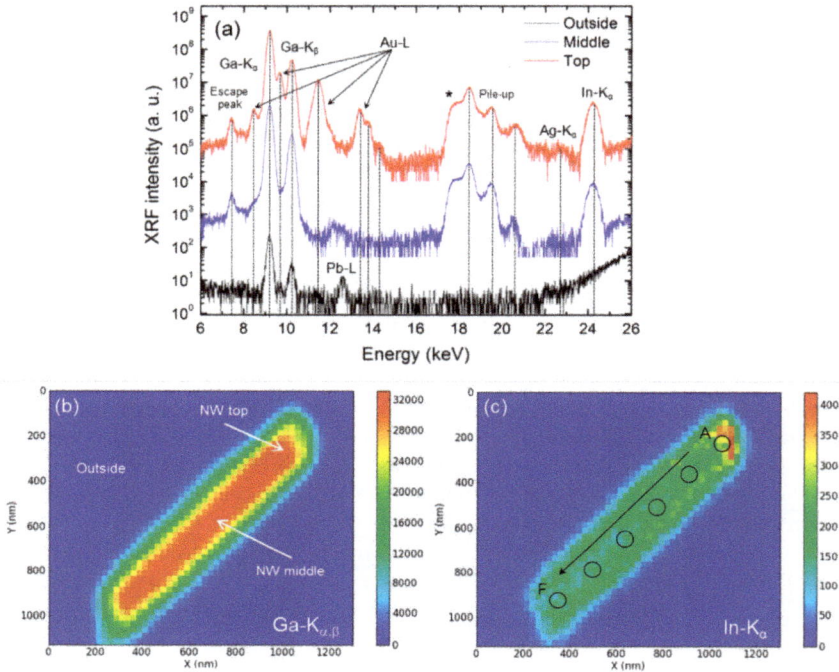

Figure 3. (**a**) Averaged X-ray flouresence (XRF) spectra from locations at the top (red) and middle (blue) of the NW, and at an outside region (black). The labels indicate the elements associated with each peak identified with PYMCA. The asterisk represents an artefact coming from the measurements. The XRF intensity color maps of Ga and In in the scanned area are shown in (**b**) and (**c**), respectively. Red (blue) color corresponds to high (low) fluorescence intensity (scale in counts). The black circles in (**c**) indicate the regions of the NW along its axis in which the In concentration has been estimated.

The distribution of Ga and In in the NW can be elucidated from the XRF intensity maps shown in Figure 3b,c. Figure 3b evidences that Ga is homogeneously distributed along the NW axis. On the other hand, the intensity of In increases from the bottom to the top end of the NW, as it is depicted in Figure 3c. The higher intensity of In at the top of the NW does not correspond to the formation of polar MQWs, as previously demonstrated by EDS. Au and Ag maps (see Figure S2) lead to the conclusion that no catalyst atoms are disseminated in the NWs. Further insight is achieved from longitudinal and radial scans of the XRF peak intensities (see Figure S3). These profiles confirm the trends that were previously found in the XRF maps: there is an InGaN shell structure with an increasing concentration of In towards the top of the NW. Ga atoms, on the other hand, are more evenly distributed.

In order to estimate the In concentration in the core-shell MQWs, the relation between the intensity of the XRF emission of an element, I_i, and its concentration, C_i, are extracted from the maps using PYMCA software. In the case of thin samples, where enhancement effects due to additional excitation of the element of interest by the characteristic radiation of other elements can be neglected, these are related by the following equation:

$$I_i \approx I_0 C_i k_i d \tag{1}$$

where I_0 is the intensity of the incoming beam, d is the sample thickness, and k_i accounts for the fluorescence yield, solid angle, and detection efficiency. The validity of Equation (1) for InGaN NWs has been extensively discussed by Gomez-Gomez et al. [17]. In a second step, the In concentration of the core-shell MQWs was estimated using the values for the thickness of the wells, barriers, and NW core obtained from the EDS analysis. Finally, the In concentration of the MQW on several points along the *c*-axis of the NW can be estimated. The obtained values are reported in Table 1 and correspond to the regions depicted in Figure 3c and are labelled from A to F. The error values are the result of the propagation of the uncertainties of the In concentration (given by PYMCA around 5%) and the well thickness (±2.7 nm). PYMCA errors are given mostly by the parameters of renormalisation used in the program; the dispersion of the XRF intensity (see Figure 3c) only contributes 0.13% and can be neglected.

Table 1. Values of the In concentration of the core-shell multiple quantum wells (MQWs) at the different positions along the axis of the NW indicated in Figure 3c.

Point	C_{In} (%)
A	11.6 ± 3.3
B	9.2 ± 2.7
C	7.4 ± 2.1
D	6.3 ± 1.8
E	5.4 ± 1.6
F	5.4 ± 1.6

3.3. Structural Properties of Individual NWs

Nano-XRD measurements can address the crystal phase and the lattice parameters of different regions across single NWs. The short distance separating the CCD and the sample allowed that three diffraction peaks were measured simultaneously. The identification of these peaks, performed with Fit2D and PYMCA programs, pointed out that they correspond to (104), (210), and (211) reflections of unstrained wurtzite GaN (we are using 3 index notation (*hkl*), which is equivalent to (*hkil*) with *i* = −(*h* + *k*)). One of these Bragg reflections is shown in log-scale in Figure 4a. The value of the GaN wurtzite reflections expected in this case were calculated with PowderCell program [27]. Neither additional reflections nor asymmetries are observed (similar results are obtained for reflections (104) and (211)). Therefore, only diffraction peaks corresponding to unstrained GaN reflections were detected studying the XRD signal of the whole NW. This fact suggests that the core-shell InGaN MQWs are completely lattice matched to the strain-free GaN core. Since the In concentration has been measured by XRF, Vergard's law can be used to calculate the lattice parameters corresponding to unstrained InGaN QWs, and from those the angular positions of the (210) and (211) Bragg reflections of InGaN. For an In concentration of 7.6%, XRD peaks should appear at $2\theta_{210}$ = 24.262° and $2\theta_{211}$ = 24.757°, which are depicted as dashed lines in Figure 4b. No distinctive XRD peak coming from the core-shell MQWs has been observed in the expected region, confirming that the core-shell MQWs are completely matched with the GaN core and therefore completely strained (compressed along *a* and *c* directions and expanded along *m*-direction).

The XRD reflexions allow one to calculate the interplanar distance d_{hkl} using the the general relation between the interplanar distances (*d*), the Miller indices (*hkl*), and lattice parameters (*a* and *c*) in the wurtzite structure,

$$d_{hkl} = \left[\frac{4}{3a^2} \left(h^2 + hk + k^2 \right) - \frac{l^2}{c^2} \right]^{-1/2} \tag{2}$$

and from them the lattice parameters *a* and *c* of the studied NWs can be measured. Symmetric reflections, such as (210), allow one to extract the lattice parameter *a* directly from Equation (2) and then use it to calculate the parameter *c* from non-symmetric reflections, such as (211). Scanning the X-ray beam along the NW axis, we can monitor any variation of the lattice parameters. Within the experimental errors, Figure 4b shows no evolution in *a* and a non-monotonous increase in *c* towards the NW top. The studied area starts at middle until the top of the NW. No signal was registered between the bottom and the middle of the NW due to the loss of the Bragg condition at imperfections on the NW. The maximum recorded variation of the axial lattice parameter *c* is 4%, which matches the increase in the In content of the NWs towards their top (see Table 1). The error associated with the lattice parameters is an average error, and it has been obtained from the difference of the value of the *c* parameter calculated from the reflections (211) and (104). The error of the calibration is the main contribution to the error of the diffraction angles, while the error of the fit performed by Fit2D program is three order of magnitude smaller than the calibration error and has not been taken into account.

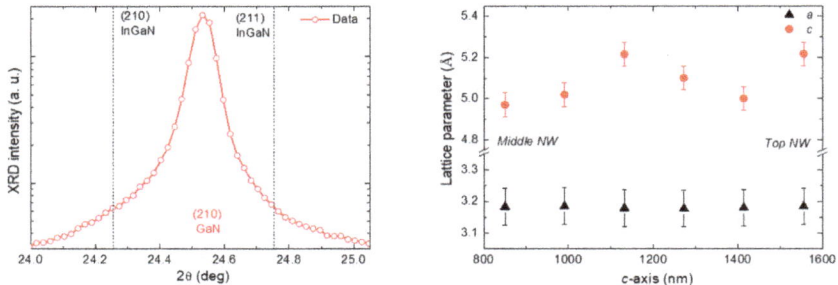

Figure 4. (**a**) XRD diffraction peak measured in the middle region of an individual NW and plotted in log-scale. The dashed lines indicate the positions of XRD peaks corresponding to unstrained GaN (210) and InGaN (210) and (211) with an average In concentration of $x = 0.076$. (**b**) Evolution of the wurtzite lattice parameters *a* and *c* along the z-axis of the NW, starting from the middle of the NW towards its top.

While previous XRD measurements were a probe of the long-range structural order, XANES measurements were performed for studying the local arrangement around Ga atoms in the NWs. The spectra were taken with a step size of 1 eV near the Ga K-edge that matches the energy resolution provided by the Si (111) double-crystal monochromator. A representative XANES of a single NW taken with the polarisation of the X-ray beam perpendicular and parallel to the axis of the NW is shown in Figure 5a,b, respectively. The XANES spectrum of the NW was compared with that measured in a high-quality *c*-oriented GaN reference layer, measured with the polarization of the X-rays either perpendicular or parallel to the wurtzite *c*-axis. The spectra of the NW match the reference ones; therefore, it can be concluded that the NW has wurtzite structure and there is no mixture of phases. Moreover, it can be argued that the axis of the NW coincides with the *c*-axis of the wurtzite structure. Similar results were obtained in the case of parallel polarisation. The small difference between the absorption edge of the NW and that of the layer is attributed to the calibration, since the measurements were carried out in different moments and therefore not exactly under the same experimental conditions.

3.4. Emission Properties of Individual NWs

The optical properties of individual and ensemble core-shell GaN/InGaN NWs were studied by PL on a statistically relevant number of individual NWs both at room and liquid He temperature. Hundreds of NWs were dispersed on a Au-patterned substrate for measuring their individual emission with the same experimental conditions. Representative PL spectra taken at 5 K with the laser light focused either at the tip or the base ends of the NWs are shown in Figure 6a. All the spectra are dominated by transitions of the GaN/InGaN core-shell MQWs with energies between 3.1–3.3 eV. The large FWHM (~34 meV) of this peak is attributed to the variation of the alloy composition along the NWs. The peak centred around 3.47 eV is attributed to emission from GaN and it is originated by the overlapping of the $D^0 X_A$ (FWHM~50 meV) and X_A emissions (FWHM~20 meV). There is a feature appearing at 3.45 eV present in all the spectra that has been attributed to an artefact coming from the gratings of the spectrometer.

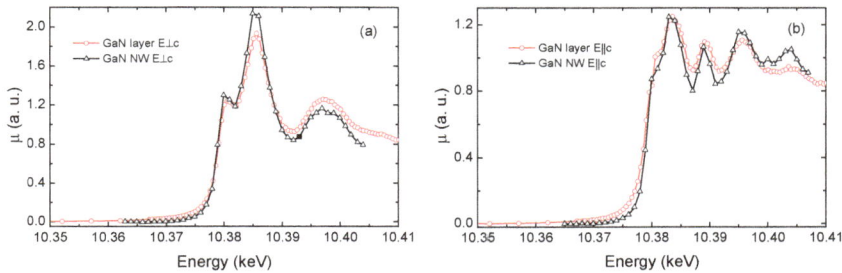

Figure 5. Comparison between X-ray absorption near edge structure (XANES) spectra of a high quality reference GaN layer (red cercles) and those of the NW (black triangles) for different X-ray beam incident angles: polarization (**a**) perpendicular and (**b**) parallel to the wurtzite *c*-axis.

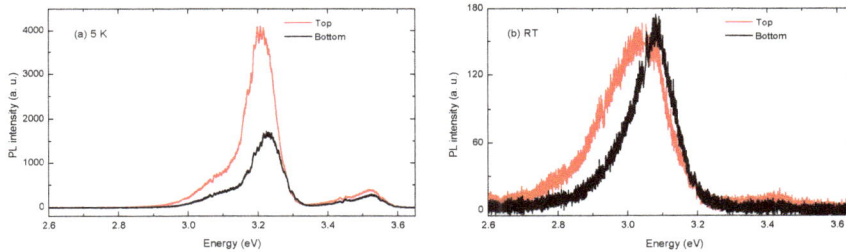

Figure 6. Representative photoluminescence (PL) spectra taken at the top (red line) and at the bottom (black line) of the NWs (**a**) at liquid He and (**b**) room temperature.

When comparing the spectra taken at the tip and the base of each NW, the PL band of InGaN MQWs increases its intensity and red-shifts from the bottom to the top of the NW. The red-shift follows from the higher In concentration towards the top end of the NW, which was evidenced by the XRF measurements. Despite the inhomogeneities present in each single NW, affecting the In concentration mostly, the PL spectra obtained from individual NWs tend to fit to a single Gaussian peak better than those recorded for NW ensembles (see Figure S4). The MQWs emission identified in macro-PL measurements often presents multi-Gaussian distributions with much greater dispersion than single NW emission. Thus, the overall inhomogeneities of the ensemble mask the real properties of the individual NWs, proving the necessity of single NW characterization.

Figure 6b shows two room temperature spectra of the same individual NW taken at its top and bottom. PL measurements as a function of the temperature allow one to calculate the relative internal quantum efficiency (IQE) of the optical transitions at room temperature. This can be done under the assumption that, at low temperature, non-radiative processes are suppressed. Thus, the relative IQE is given by the ratio between the PL integrated intensity at room temperature and low temperatures. At room temperature, IQE of the tip and of the base of the core-shell InGaN MQWs is around 20% and 15%, respectively. On the other hand, the IQE of the core is less than 1%, as it is expected for the absence of quantum confinement.

3.5. Theoretical Simulations

In order to get further insight into the optical properties of the investigated non-polar InGaN/GaN QW NWs, a theoretical model has been implemented. First of all, XRD results point out that strain fields are present in the heterostructures that will affect strongly their optical response. The strain field calculations in 3D systems tend to be time- and CPU-expensive. Thus, it is preferable to reduce the problem to a 2D one by assuming that the strain depends only on the in-plane coordinates (x, y) and is invariant along z. This approximation gives good results in the case of a finite but long 3D system, which perfectly matches the geometry of the NWs studied here [28]. The numerical calculations have been performed using the COMSOL-Multiphysics software platform [29] for a GaN/InGaN MQW NW with averaged values of the thickness of the wells and the barriers (4.5 and 5.4 nm, respectively, resulting from EDS measurements) and In concentration of 10% (extracted from XRF data). It is assumed that the only source of strain is given by the lattice-mismatch between InGaN and GaN.

A cross-sectional contour map of the angular strain component (ε in the chosen cylindrical coordinates) is represented in Figure 7a. The GaN core is essentially relaxed (all strain components are in the range 0–0.2%). Inside the InGaN QWs, $\varepsilon_{\phi\phi} \sim -0.8\%$ (similar values are obtained for ε_{zz}, see Figure S5) and indicates that the three QWs are under compressive strain due to the larger lattice parameter of InGaN compared to that of GaN. The radial component is tensile strained as a result of the compression along the angular and c-axis directions ($\varepsilon_{rr} \sim 0.4\%$). The distribution of the strain is homogeneous, except for the corners where, due to their particular shape, the strain field has a more complicated behavior. These results confirm the observations made by the XRD: GaN/InGaN MQWs grow strained and matched to the GaN material. Figure 7b presents the evolution of the angular strain component along the diameter of the NW.

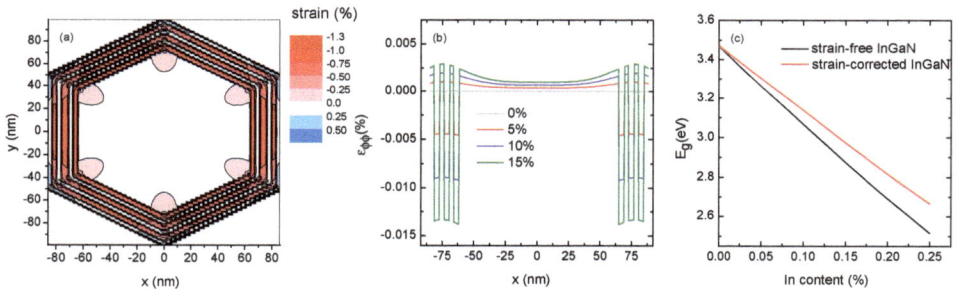

Figure 7. (**a**) Cross-sectional map of the angular strain component (ε) for InGaN MQWs with 10% In concentration. (**b**) Evolution of the angular strain component along the x-direction ($y = 0$) for different In concentrations. (**c**) Band gap energy of strain-free and strained InGaN for different In concentrations.

The effect of changing the In concentration has been modeled from 0 up to 15% and, within this range, the average value of the strain components changes linearly. Using the obtained strain field distribution, the electronic band structure of strained InGaN can be calculated. Figure 7c shows the trend of InGaN bandgap energy (E_g) at low temperatures for different In concentrations. The strain-free values of the InGaN bandgap are also plotted for comparison. As expected, due to the compressive character of the strain field, strained InGaN has band gap energies higher than those of relaxed material. The values of In concentration in the GaN/InGaN core-shell MQWs, obtained from the XRF maps, are in the range of 5–12%, and the corresponding bandgap energies are in the 3.05–3.29 eV range. This result is in very good agreement with the PL measurements. The large FWHM (~34 meV) of the emission is also consistent with the variation of In concentration along the NW axis.

4. Conclusions

The elemental, structural, and optical characterization of GaN/InGaN core-shell NWs requires a battery of complementary and non-destructive techniques. Due to the complexity of these nanostructures, including inhomogeneities and strain at a nanometer length scale, the combination of more than one technique, in such a manner that the strengths of one complement the weaknesses of others, is indispensable. The combination of XRF and EDS gives a reliable picture of the dimensions of the QWs and their alloy composition. Variations in In content along the axial direction are present in all NWs, and the upper tips tend to be In richer than the bottom ends. XANES and XRD probe local and long-range order in the heterostructures. Knowing the In concentration in the QWs, their strain state can be inferred from XRD spectra. Finally, photoluminescence spectra of individual NWs can be interpreted and modeled theoretically. Fluctuations in the In content can explain the variations in the emission energy and the broadening of the emission peaks.

Supplementary Materials: The following are available online at http://www.mdpi.com/2079-4991/9/5/691/s1, Figure S1: EDS linear scan along the NW axis, Figure S2: XRF maps of Au-L and Ag-Kα, Figure S3: Linear scans of In and Ga XRF peaks performed along the NW axis and NW diameter, Figure S4: PL spectrum of an ensemble of NWs, Figure S5: Contour plots of the strain components on the NW cross-section.

Author Contributions: Conceptualization, N.G., G.M.-C., and A.G.-C.; Methodology, J.S.-R., C.G., N.G., G.M.-C., and A.G.-C.; Validation, E.S., H.T.M., A.G.-C., G.M.-C., and N.G.; Formal Analysis, E.S., J.S.-R., G.M.-C., H.T.M., and A.G.-C.; Investigation, E.S., B.F., and H.T.M.; Resources, H.B., C.G., and M.H.; Writing—Original Draft Preparation, E.S.; Writing—Review and Editing, N.G., G.M.-C., J.S.-R., and A.G.-C.; Visualization, E.S. and H.T.M.; Supervision, N.G., G.M.-C, and A.G.-C.; Project Administration, N.G., C.G., and A.C.; Funding Acquisition, A.C. and N.G.

Funding: This research was funded by the European Union Seventh Framework Programme, grant number 265073; and by the Spanish MICINN and the FEDER funds of the European Union through the project ENE2016-79282-C5-3-R.

Acknowledgments: The authors thank the European Synchrotron Radiation Facility (ESRF) for the beamtime allocated at beamline ID22-NI.

Conflicts of Interest: The authors declare no conflict of interest.

References

1. Li, S.; Waag, A. GaN based nanorods for solid state lighting. *J. Appl. Phys.* **2012**, *111*, 071101. [CrossRef]
2. Yan, L.; Jahangir, S.; Wight, S.A.; Nikoobakht, B.; Bhattacharya, P.; Millunchick, J.M. Structural and Optical Properties of Disc-in-Wire InGaN/GaN LEDs. *Nano Lett.* **2015**, *15*, 1535. [CrossRef] [PubMed]
3. Lu, W.; Lieber, C.M. Semiconductor nanowires. *J. Phys. D Appl. Phys.* **2006**, *39*, R387. [CrossRef]
4. Kuykendall, T.; Ulrich, P.; Aloni, S.; Young, P. Complete composition tunability of InGaN nanowires using a combinatorial approach. *Nat. Mater.* **2007**, *6*, 951. [CrossRef]
5. Qian, F.; Gradečak, S.; Li, Y.; Wen, C.-Y.; Lieber, C.M. Core/multishell nanowire heterostructures as multicolor, high-efficiency light-emitting diodes. *Nano Lett.* **2005**, *5*, 2287. [CrossRef] [PubMed]
6. Ra, Y.-H.; Navamathavam, R.; Kang, S.; Lee, C.-R. Different characteristics of InGaN/GaN multiple quantum well heterostructures grown on m- and r-planes of a single n-GaN nanowire using metalorganic chemical vapor deposition. *J. Mater. Chem. C* **2014**, *2*, 2692. [CrossRef]
7. Ra, Y.-H.; Navamathavam, R.; Yoo, H.-I.; Lee, C.-R. Single Nanowire Light-Emitting Diodes Using Uniaxial and Coaxial InGaN/GaN Multiple Quantum Wells Synthesized by Metalorganic Chemical Vapor Deposition. *Nano Lett.* **2014**, *14*, 1537. [CrossRef]
8. Koester, R.; Sager, D.; Quitsch, W.-A.; Pfingsten, O.; Poloczek, A.; Blumenthal, S.; Keller, G.; Prost, W.; Bacher, G.; Tegude, F.-J. High-Speed GaN/GaInN Nanowire Array Light-Emitting Diode on Silicon(111). *Nano Lett.* **2015**, *15*, 2318. [CrossRef] [PubMed]
9. Li, C.; Wright, J.B.; Liu, S.; Lu, P.; Figiel, J.J.; Leung, B.; Chow, W.W.; Brener, I.; Koleske, D.D.; Luk, T.-S.; et al. Nonpolar InGaN/GaN Core-Shell Single Nanowire Lasers. *Nano Lett.* **2017**, *17*, 1049–1055. [CrossRef]
10. Zhang, H.; Guan, N.; Piazza, V.; Kapoor, A.; Bougerol, C.; Julien, F.H.; Babichev, A.V.; Cavassilas, N.; Bescond, M.; Michelini, F.; et al. Comprehensive analyses of core-shell InGaN/GaN single nanowire photodiodes. *J. Phys. D Appl. Phys.* **2017**, *50*, 484001. [CrossRef]
11. Koester, R.; Hwang, J.-S.; Salomon, D.; Chen, X.; Bougerol, C.; Barnes, J.-P.; Dang, D.L.; Rigutti, L.; Bugallo, A.D.; Jacopin, G.; et al. M-Plane Core-Shell InGaN/GaN Multiple-Quantum-Wells on GaN Wires for Electroluminescent Devices. *Nano Lett.* **2011**, *11*, 4839. [CrossRef]

12. Yeh, T.-W.; Lin, Y.-T.; Stewart, L.W.; Dapkus, P.D.; Sarkissian, R.; O'Brien, J.D.; Ahn, B.; Nutt, S.R. InGaN/GaN Multiple Quantum Wells Grown on Nonpolar Facets of Vertical GaN Nanorod Arrays. *Nano Lett.* **2012**, *12*, 3257. [CrossRef] [PubMed]

13. Riley, J.R.; Padalkar, S.; Li, Q.; Lu, P.; Koleske, D.D.; Wierer, J.J.; Wang, G.T.; Lauhon, L.J. Three-Dimensional Mapping of Quantum Wells in a GaN/InGaN Core-Shell Nanowire Light-Emitting Diode Array. *Nano Lett.* **2013**, *13*, 4317. [CrossRef]

14. Ra, Y.-H.; Navamathavan, R.; Park, J.-H.; Lee, C.-R. Coaxial InxGa1-xN/GaN Multiple Quantum Well Nanowire Arrays on Si(111) Substrate for High-Performance Light-Emitting Diodes. *Nano Lett.* **2013**, *13*, 3506–3516. [CrossRef]

15. Kim, J.-H.; Ko, Y.-H.; Cho, J.-H.; Gong, S.-H.; Ko, S.-M.; Cho, Y.-H. Toward highly radiative white light emitting nanostructures: A new approach to dislocation-eliminated GaN/InGaN core-shell nanostructures with a negligible polarization field. *Nanoscale* **2014**, *6*, 14213–14220. [CrossRef] [PubMed]

16. Segura-Ruiz, J.; Martinez-Criado, G.; Sans, J.A.; Tucoulou, R.; Cloetens, P.; Snigireva, I.; Denker, C.; Malindretos, J.; Rizzi, A.; Gómez-Gómez, M.; et al. Direct observation of elemental segregation in InGaN nanowires by X-ray nanoprobe. *Phys. Status Solidi (RRL)* **2011**, *5*, 95–97. [CrossRef]

17. Gómez-Gómez, M.; Garro, N.; Segura-Ruiz, J.; Martinez-Criado, G.; Chu, M.H.; Cantarero, A.; Mengistu, H.T.; García-Cristóbal, A.; Murcia-Mascarós, S.; Denker, C.; et al. Spontaneous core-shell elemental distribution in In-rich InxGa1-xN nanowires grown by molecular beam epitaxy. *Nanotechnology* **2013**, *25*, 075705. [CrossRef] [PubMed]

18. Segura-Ruiz, J.; Martinez-Criado, G.; Chu, M.H.; Denker, C.; Malindretos, J.; Rizzi, A. Synchrotron nanoimaging of single In-rich InGaN nanowires. *J. Appl. Phys.* **2013**, *113*, 136511. [CrossRef]

19. Segura-Ruiz, J.; Martinez-Criado, G.; Chu, M.H.; Denker, C.; Malindretos, J.; Rizzi, A. Phase Separation in Single InxGa1-xN Nanowires Revealed through a Hard X-ray Synchrotron Nanoprobe. *Nano Lett.* **2014**, *14*, 1300–1305. [CrossRef] [PubMed]

20. Stankevič, T.; Dzhigaev, D.; Bi, Z.; Rose, M.; Shabalin, A.; Reinhardt, J.; Mikkelsen, A.; Samuelson, L.; Falkenberg, G.; Vartanyants, I.A.; et al. Strain mapping in an InGaN/GaN nanowire using a nano-focused x-ray beam. *Appl. Phys. Lett.* **2015**, *107*, 103101. [CrossRef]

21. Krause, T.; Hanke, M.; Nicolai, L.; Cheng, Z.; Niehle, M.; Trampert, A.; Kahnt, M.; Falkenberg, G.; Schroer, C.G.; Hartmann, J.; et al. Structure and Composition of Isolated Core-Shell (In,Ga)N/GaN Rods Based on Nanofocus X-Ray Diffraction and Scanning Transmission Electron Microscopy. *Phys. Rev. Appl.* **2017**, *7*, 024033. [CrossRef]

22. Al Hassan, A.; Lewis, R.B.; Küpers, H.; Lin, W.-H.; Bahrami, D.; Krause, T.; Solomon, D.; Tahroaui, A.; Hanke, M.; Geelhaar, L.; et al. Determination of indium content of GaAs/(In,Ga)As/(GaAs) core-shell(-shell) nanowires by x-ray diffraction and nano x-ray fluorescence. *Phys. Rev. Mater.* **2018**, *2*, 014604. [CrossRef]

23. Solé, V.A.; Papillon, E.; Cotte, M.; Water, P.; Susini, J. A multiplatform code for the analysis of energy-dispersive X-ray fluorescence spectra. *J. Spectrochim. Acta B* **2007**, *62*, 63–68. [CrossRef]

24. Hammersley, A.P.; Svensson, S.O.; Hanfland, M.; Fitch, A.N.; Husermann, D. Two-dimensional detector software: From real detector to idealised image or two-theta scan. *High Press. Res.* **1996**, *14*, 235–248. [CrossRef]

25. Newville, M.J. EXAFS analysis using FEFF and FEFFIT. *J. Synchrotron Radiat.* **2001**, *8*, 96–100. [CrossRef]

26. Mata, R.; Cros, A.; Hestroffer, K.; Daudin, B. Surface optical phonon modes in GaN nanowire arrays: Dependence on nanowire density and diameter. *Phys. Rev. B* **2012**, *85*, 035322. [CrossRef]

27. Available online: http://www.ccp14.ac.uk/tutorial/powdcell (accessed on 1 June 2016).

28. Mengistu, H.T.; Garcia-Cristóbal, A. The generalized plane piezoelectric problem: Theoretical formulation and application to heterostructure nanowires. *Int. J. Solids Struct.* **2016**, *100*, 257–269. [CrossRef]

29. Available online: http://www.comsol.com/comsol-multiphysics (accessed on 1 June 2016).

nanomaterials

MDPI

Article

Extracting the Dynamic Magnetic Contrast in Time-Resolved X-ray Transmission Microscopy

Taddäus Schaffers [1,*], Thomas Feggeler [2], Santa Pile [1], Ralf Meckenstock [2], Martin Buchner [1], Detlef Spoddig [2], Verena Ney [1], Michael Farle [2], Heiko Wende [2], Sebastian Wintz [3,4], Markus Weigand [5], Hendrik Ohldag [6], Katharina Ollefs [2] and Andreas Ney [1,*]

[1] Institute of Semiconductor and Solid State Physics, Johannes Kepler University Linz, 4040 Linz, Austria
[2] Faculty of Physics and Center for Nanointegration Duisburg-Essen (CENIDE), University of Duisburg-Essen, 47057 Duisburg, Germany
[3] Paul Scherrer Institut, 5232 Villigen PSI, Switzerland
[4] Helmholtz-Zentrum Dresden-Rossendorf, 01328 Dresden, Germany
[5] Max-Planck-Institut für Intelligente Systeme, 70569 Stuttgart, Germany
[6] Stanford Synchrotron Radiation Laboratory, SLAC National Accelerator Laboratory, Menlo Park, CA 94025, USA
* Correspondence: taddaeus.schaffers@jku.at (T.S.); andreas.ney@jku.at (A.N.)

Received: 28 May 2019; Accepted: 24 June 2019; Published: 28 June 2019

Abstract: Using a time-resolved detection scheme in scanning transmission X-ray microscopy (STXM), we measured element resolved ferromagnetic resonance (FMR) at microwave frequencies up to 10 GHz and a spatial resolution down to 20 nm at two different synchrotrons. We present different methods to separate the contribution of the background from the dynamic magnetic contrast based on the X-ray magnetic circular dichroism (XMCD) effect. The relative phase between the GHz microwave excitation and the X-ray pulses generated by the synchrotron, as well as the opening angle of the precession at FMR can be quantified. A detailed analysis for homogeneous and inhomogeneous magnetic excitations demonstrates that the dynamic contrast indeed behaves as the usual XMCD effect. The dynamic magnetic contrast in time-resolved STXM has the potential be a powerful tool to study the linear and nonlinear, magnetic excitations in magnetic micro- and nano-structures with unique spatial-temporal resolution in combination with element selectivity.

Keywords: ferromagnetic resonance; X-ray magnetic circular dichroism; scanning transmission X-ray microscopy

---

1. Introduction

In spintronics and magnonics, it is important to understand the magnetization dynamics on the micro- and nano-scale e.g., to be able to control the propagation of spin waves. A well-established technique to measure the dynamic magnetic behavior of a system is ferromagnetic resonance (FMR). However, classical resonator based FMR measurements are not able to detect single micro- or nano-sized objects due to their detection limit of around 10^{11} spins [1]. This sensitivity limit has been overcome in recent years by the development of lithographically fabricated micro-resonators [2], which are capable of measuring down to 10^6 spins [3], corresponding to a single Fe-nanocube with dimensions of $30 \times 30 \times 30$ nm^3. Due to the lack of spatial resolution below the diameter of the micro-resonator of typically a few tens of microns, it is impossible to separate the FMR signal of a single nano-particle from the resonance signal of the whole ensemble during the homogeneous excitation of the micro-resonator cavity.

To facilitate spatial resolution, other measurement techniques have been combined with FMR excitation in order to measure a single nano-sized object in an ensemble. These measurement techniques include but are not limited to: magneto optic Kerr effect (MOKE) [4], Brillouin light scattering (BLS) [5], magnetic force microscopy (MFM) [6], scanning thermal microscopy (SThM) [7], scanning electron microscopy with polarization analysis (SEMPA) [8], and X-ray photoemission electron microscopy (X-PEEM) [9]. For most of these measurement techniques, it is not possible to measure with element selectivity (MOKE, BLS, MFM, SThM and SEMPA), while other measurement techniques like X-PEEM can only probe the surface of the sample with element selectivity. In recent years, the X-ray magnetic circular dichroism (XMCD) [10–12] effect has been combined with FMR in order to probe the dynamic magnetic excitation, the so-called X-ray detected ferromagnetic resonance (XFMR) [13,14], utilizing the element selectivity of the X-rays. A spatial resolution of down to 20 nm can be achieved by using a scanning transmission X-ray microscope (STXM).

By combining the micro-resonator FMR with STXM (STXM-FMR) within a synchronization scheme for the exciting microwaves and the probing X-ray photons of the synchrotron, it is possible to detect FMR with a high temporal (ps-regime) as well as spatial resolution (nm regime) [7,14–16]. Combining these features, STXM-FMR measurements bare the potential to significantly deepen our understanding of the dynamic magnetization of ferromagnetic heterosystems containing different chemical elements [17] as well as non-ferromagnets with induced magnetization [18].

In order to be able to draw valid conclusions from the dynamic magnetic contrast in STXM-FMR, it is necessary to perform a range of control-experiments in the first place as well as testing the robustness of the evaluation of the raw data to establish that STXM-FMR indeed provides significant information about the dynamic magnetic behavior of a given magnetic specimen based on the XMCD effect. In this paper, a range of control experiments will be presented as well as a detailed analysis of the separation of the true magnetic contrast from background effects. The obtained results allow for reliably image homogeneous and inhomogeneous magnetic excitations in magnetic micro-stuctures with very high spatio-temporal resolution. Furthermore, it is possible to obtain quantitative information about the local precession angle in FMR and its relative phase within a given STXM-FMR experiment.

2. Experimental Details

The magnetic specimen is placed inside a micro-resonator and microwaves are used to excite the FMR. The micro-resonator is fabricated on a 200 nm thick, $250 \times 250\ \mu m^2$ large silicon nitride membrane suspended by a $5 \times 10\ mm^2$ silicon frame of high resistivity. In a first step, the magnetic specimen is fabricated on the SiN-membrane using electron beam lithography (EBL). Two different designs for the magnetic specimen were made. The first one consists of two perpendicular permalloy (Py) stripes with dimensions of $5 \times 1 \times 0.03\ \mu m^3$ (see Figure 1b), which are deposited using magnetron sputtering at room temperature and capped with aluminum. The second sample system is a combination of a Py disk with a Co stripe. For this in a first step, a Py disk with a diameter of 2.6 µm and a thickness of 30 nm is fabricated. With a second EBL step, a Co stripe with lateral dimensions of $2 \times 0.6\ \mu m^2$ and a thickness of 30 nm is placed on top of the Py disk (see Figure 1c). Finally, the micro-resonator is patterned around the magnetic specimen using optical lithography (OL), leaving the sample inside the Ω shaped resonator loop in Figure 1a. The gold used to produce the micro-resonator has a thickness of 600 nm and an additional 5 nm of titanium is used as an adhesion layer. Both materials are deposited by thermal evaporation.

By utilizing the STXM, it is possible to measure with a spatial resolution of 35 nm at SSRL and 20 nm at the MAXYMUS beamline at BESSY II, achieved by focusing the X-rays onto the magnetic specimen using a zone-plate. To measure the STXM-FMR at the Stanford Synchrotron Radiation Lightsource (SSRL), beamline 13.1, the FMR excitation needs to be synchronized to the bunch frequency of the synchrotron, thus enabling us to measure the FMR precession with time resolution. The stroboscopic time resolution for the STXM-FMR measurement is achieved by phase locking the GHz microwave frequency to the 476.315 MHz bunch frequency of the SSRL synchrotron.

Furthermore, a PIN-diode was installed to switch the microwave on and off with the synchrotron revolving frequency of 1.28 MHz. This comparison of X-ray transmission detected with and without applied microwave power allows for detecting very small changes in the X-ray transmission as a result of the magnetization precession. A fast avalanche photo diode (APD) detects the transmitted X-ray photons behind the sample. The APD signal is finally stored in 12 different channels. Each of these channels corresponds to the signal of a specific group of X-ray pulses. The first six channels are used for the APD signal of the transmitted X-rays with applied microwave, while the second six channels are used to measure the X-ray transmission without applied microwave. For the first six channels, the magnetization inside the sample is precessing, while the magnetization is static for the second six channels. Each of these channels corresponds to a specific relative phase of the FMR precession with respect to the microwave excitation. The latter non-precessing channels are crucial for eliminating the influence of the filling pattern of the bunch train of electron buckets on the resulting STXM images. The six different channels correspond to six specific bunches that are phase shifted each by 60°, with respect to the microwave frequency of up to 9.6 GHz. Therefore, the six phases correspond to time resolved snap-shot images which are separated by 17.4 ps, and comprise one full precession cycle of the magnetization. One should note that each X-ray flash has a pulse duration of 50 ps, which fundamentally limits the attainable temporal resolution. Additional information regarding the synchronization scheme and the X-ray microscope can be found in [7,15].

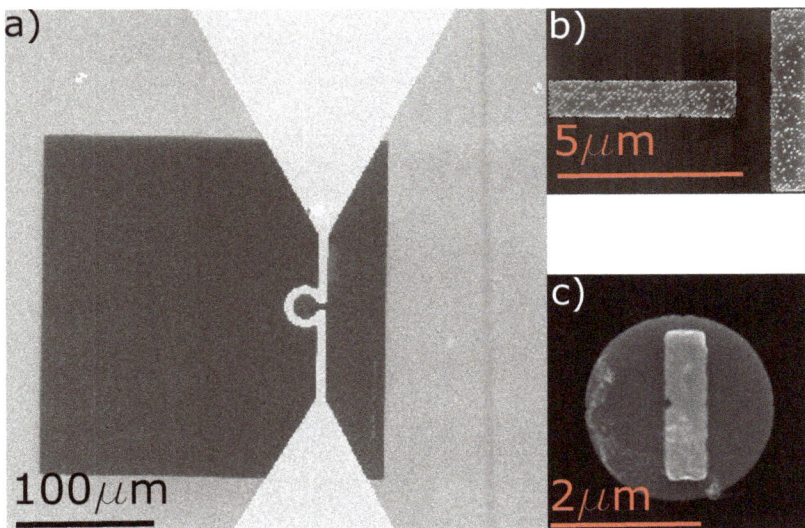

Figure 1. (**a**) Scanning electron microscope (SEM) image of the strip-line resonator on top of a SiN-membrane. For this work, two different sample systems were chosen. In (**b**), two perpendicular Py stripes ("T"-sample) are shown, while in (**c**) the Py-Co disk stripe sample can be seen.

A similar approach for measuring the dynamic magnetization in an FMR experiment with spatial and temporal resolution has been implemented at the Maxymus endstation at BESSY II [19,20]. There are two moderate differences with respect to the SSRL experiment: one is that the BESSY II operation frequency is appx. 500 MHz, corresponding to a repetition period of the probing X-ray flashes of 2 ns. Secondly, the signal is recorded only for the microwave on (precessing) case. Therefore, on the one hand, it is not necessary to excite the sample at direct higher harmonic frequencies of the synchrotron and thus the exciting frequencies can be chosen more freely to f = 500 MHz·M/N, depending on the number of detection channels used (N) and a selectable integer multiplier M [16]. Here, N for most cases is also equal to the number of simultaneously acquired excitation phases (not limited to 6). Since the non-precessing magnetization (microwave off) cannot be used as a baseline for comparison only the transmitted intensity ratio of each channel with respect to the temporal average $I(t)/<I>t$ can be used for extracting the dynamic magnetic contrast associated with the precession of magnetization. In addition, for each channel on average, all bunches contribute equally so that the filling pattern of the ring is averaged out.

3. Contrast Mechanism

For a better understanding of the measured STXM-FMR data, we briefly discuss the underlying physical effect, which yields the dynamic magnetic contrast images.

3.1. X-ray Absorption

The transmission of electromagnetic radiation through any material is described by Beer–Lambert's law [21]. In an STXM, the transmitted X-ray intensity I is detected in normal incidence. This transmitted intensity is controlled by the composite X-ray absorption (XA) coefficient of the entire sample (magnetic specimen and SiN-membrane). Tuning the photon energy to any characteristic core-level excitation results in the well-known element selectivity of XA measurements. However, the above-mentioned law only considers a single layer system. In an STXM-FMR experiment, the sample consists at least of a two layers with distinct properties, since any magnetic specimen is supported by the SiN-membrane through which the X-rays need to be transmitted as well. In order to include this second layer, a Beer-Lambert's law needs to be modified [21]:

$$I_{s/m} = I_0 e^{-(\mu_s t_s + \mu_m t_m)}, \tag{1}$$

where t_s, t_m are the thicknesses and μ_s, μ_m are the absorption coefficients of the magnetic specimen and SiN-membrane, respectively, and I_0 is the incoming intensity. In any sample, one can find areas where the X-rays only transmit through the SiN-membrane while in other regions the X-rays are transmitted through the SiN-membrane plus the magnetic specimen, which can also consist of more than one layer that would be added to the exponent in Equation (1). Therefore, the dynamic magnetic contrast can be separated from the background transmission through the SiN-membrane by defining the respective regions of interests (RoI) from the individual, time-averaged z-contrast STXM-FMR images like in Figure 2b or Figure 3.

3.2. XMCD Effect in STXM-FMR

It is well-known that for circular polarized X-rays at the $L_{3/2}$-edges of ferromagnetic transition metals the XA coefficient μ depends on the relative orientation of the magnetization M and the polarization vector σ^+ and σ^- of the synchrotron light, respectively, called the XMCD effect [11,12]:

$$I(\sigma^{+/-}) = I_0 e^{-\mu^{+/-} \cdot t}, \tag{2}$$

where $\mu^{+/-}$ is the absorption coefficient with the magnetization M parallel/antiparallel to the polarization vector $\sigma^{+/-}$. The XMCD in XA spectroscopy is commonly defined as the difference in absorption spectra between parallel and antiparallel orientation, i.e., for XMCD in transmission geometry $(\mu^+ - \mu^-) \cdot t = \Delta\mu \cdot t$. In the following, the usual spectroscopic definition of the XMCD effect will be used rather than the so-called XMCD asymmetry $(I(\sigma^+) - I(\sigma^-))/(I(\sigma^+) + I(\sigma^-))$ which can also be found throughout the literature. However, other than in normal XA spectroscopy in STXM-FMR, the images are only taken for a fixed photon energy for which the XMCD signal is maximal.

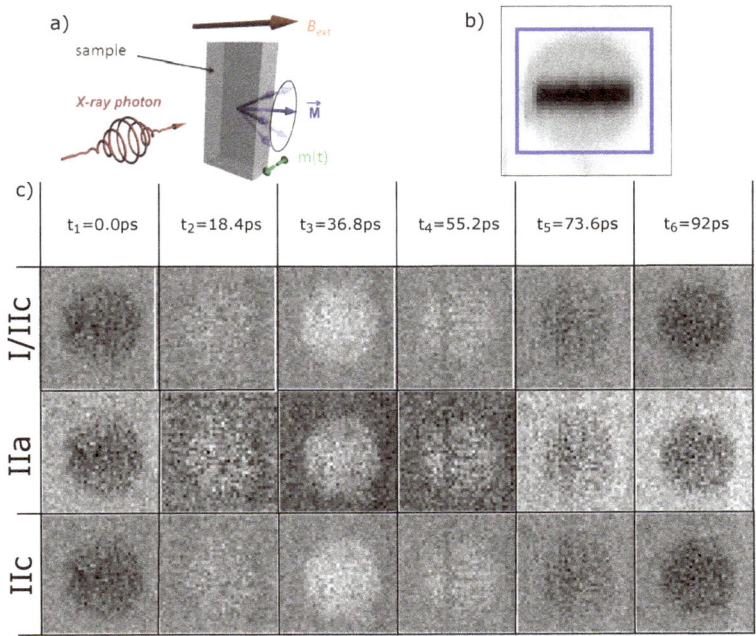

Figure 2. (**a**) sketch of the experimental geometry. The polarized photons hit the sample at normal incidence; the static magnetization is oriented parallel to the external magnetic field B_{ext}. At ferromagnetic resonance, M precesses and a time-dependent out-of-plane component $m(t)$ exists; (**b**) the chemical contrast image of the disk stripe sample measured at the Ni-L_3-edge. Red and blue frames define different regions of interest; (**c**) representation of different evaluation methods for the six phases of the magnetization precession: in the absorption coefficient, the difference is shown obtaining the background corrected microwave on and off measurement. IIa is the ratio between the not background corrected microwave on and off measurement. Applying a background correction to IIa, the images labelled IIb are generated—for details, see text.

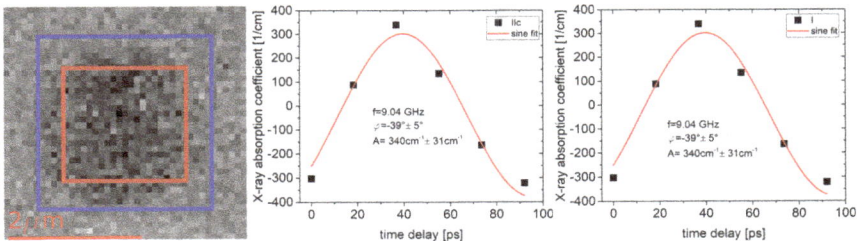

Figure 3. By averaging over the sample in the six different phases for evaluation method IIc (left side) and I (right side) from Figure 2, one can obtain the curves shown. Both were fitted with a sine function due to the sinusoidal behavior of the exciting microwave. The frequency of this sinusoidal was given by the microwave frequency applied to the system which in this case was 9.04 GHz.

For a single STXM-FMR image, neither the circular polarization nor the direction of the magnetization is reversed. This is due to the fact that the static magnetizatioj M is parallel to the external static magnetic field B_{ext} which in turn is perpendicular to the incident light, i.e., a transverse geometry with X-ray beam normal to the surface and M in the sample plane is chosen—see sketch in Figure 2a.

Therefore, only a small, dynamic out of plane component $m(t)$ of the precessing magnetization is accessible by the XMCD effect. At the SSRL, the difference between microwave on and off is therefore the difference between the precessing (= finite $m(t)$ out-of-plane) and non-precessing case ($m(t) = 0$ for all t) is recorded via the XMCD effect. In contrast, in the detection scheme at BESSY II, the average over all images corresponding to one full cycle of precession is taken, in which the dynamic magnetization component averages out and is subsequently subtracted from each individual phase. Both methods are equivalent in the sense that the XMCD effect only senses the finite out-of-plane component of the precessing magnetization $m(t)$ which requires the time-resolved detention scheme which records snapshot images of $m(t)$ at different points in time throughout a full precession cycle—for details, see [15].

4. Analysis of STXM-FMR Measurements

In light of the preceding discussion, evaluation methods for the extraction of quantitative information from the STXM-FMR data will be presented, with special attention to how to extract the dynamic contribution $m(t)$ of the magnetic specimen. Additionally, it is possible to quantify the opening angle of the magnetization precession in FMR directly from the change in absorption coefficient during a full precession cycle.

4.1. Raw Data Treatment

To eliminate the second absorption coefficient μ_m in Equation (1), the raw data needs to be corrected by the SiN-membrane background. Thus, the absorption coefficient of the magnetic specimen alone can be investigated. For that, we average the transmission signal over the area of only the SiN-membrane for each of the 12 images (six phases with microwave on I_m^{on} and six phases with microwave off I_m^{off}) separately. This corresponds to the area outside the blue box in Figure 2b. Each individual image is then divided by its respective averaged transmission value of the SiN-membrane. The resulting transmission I^{on} and I^{off} then contains exclusively the information about the absorption coefficient μ_s of the magnetic specimen:

$$I_s^{on} = \frac{I_{s/m}^{on}}{I_m^{on}} = e^{-\mu_s^{on} \cdot t_s} \qquad I_s^{off} = \frac{I_{s/m}^{off}}{I_m^{off}} = e^{-\mu_s^{off} \cdot t_s}. \tag{3}$$

This procedure also eliminates random fluctuations of the incoming intensity I_0 for each phase. Subsequently, the dynamic magnetic contrast is derived by taking the natural logarithm of the ratio of the precessing (microwave on) versus non-precessing (microwave off) case to obtain $\Delta\mu$ corresponding to the difference in absorption coefficient equivalent to the usual definition of the XMCD effect:

$$ln\left(\frac{I_s^{on}}{I_s^{off}}\right) = (\mu_s^{off} - \mu_s^{on}) \cdot t = \Delta\mu \cdot t, \tag{4}$$

where t is the thickness of the magnetic specimen. The resulting dynamic magnetic contrast $\Delta\mu \cdot t$ of the Py disk recorded at the Ni L_3-edge at 9.04 GHz is shown for all six phases in Figure 2c, row I/IIc. It is clearly visible that only the contrast of the Py disk reverses during a full measurement cycle representing the perpendicular component of the high-frequency magnetization $m(t)$, while the background stays constant.

However, one can change the sequence of extracting $\Delta\mu \cdot t$ and take a closer look at each individual step. First, the ratio of microwave on and microwave off is taken and all six phases are displayed in Figure 2 row IIa. It is obvious that the background corresponding to the SiN-membrane oscillates as well, which will be discussed further below. In a second step, the influence of the oscillating background is corrected as mentioned before by dividing each phase with the respective averaged transmission of the SiN-membrane derived outside the blue area. The resulting six phases are shown in Figure 2c, row IIb and already compare well with the full analysis of I, revealing no visible background oscillation.

However, the images of IIb do not directly reflect the numerical values of $\Delta\mu \cdot t$. Taking the natural logarithm of IIb, one obtains the numerically identical six phase images as shown in Figure 2c, row I/IIc. A direct comparison between IIb and I/IIc reveals that the qualitative behavior of the dynamic magnetic contrast is identical. However, only IIc shall be mathematically equivalent to the full analysis in I. Both evaluations depend on the selection of the RoI from which the background of the SiN-membrane is derived, see Figure 3 below.

To verify if the sequence of the evaluation steps indeed yield the same results, the quantitative outcome of methods I and IIc are compared in Figure 3. In the STXM-FMR image, the area outside the blue box defines the RoI used for determining the background of the SiN-membrane. The red box indicates the RoI which is used for determining the average $\Delta\mu \cdot t$ of the magnetic specimen. To derive the absorption coefficient $\Delta\mu$ at Ni L_3-edge, the resulting averaged value has to be divided by the effective thickness $t = 24$ nm, since the Py film is 30 nm thick and contains 80% nickel; note that the non-resonant XA of the iron can be excluded due to the ratio between the measurements with and without applied microwave power. The two panels show the averaged values (symbols) of the six phases for method I (right) and IIc (left) reflecting the dynamic magnetic contrast of the homogeneous excitation, i.e., uniform mode of the Py disk. The sine fits (solid lines) are done for the fixed frequency of the exciting microwave of 9.04 GHz while amplitude A and phase φ are fitting parameters. Indeed, both methods reveal identical numerical values for $A = \Delta\mu$ and φ of (340 ± 31) cm^{-1} and $-39° \pm 5°$, respectively. The quantitative numerical values will be discussed in the following.

4.2. Precession Angle

The first quantity that can be extracted from a STXM-FMR experiment is the amplitude A corresponding to the dynamic magnetic contrast $\Delta\mu$. For a known thickness t of the magnetic specimen, one can extract the opening angle θ of the precessing magnetization. Other than for the phase φ, the amplitude A and thus $\Delta\mu$ can be compared between different samples. For that, a usual XMCD experiment is carried out on a specimen of known thickness d where $\Delta\mu_{XMCD}$ is derived as the difference in absorption with the magnetization fully parallel and antiparallel to the k vector of the X-rays, yielding $\Delta\mu_{abs} = \Delta\mu_{XMCD}/d$. One should keep in mind that in an XMCD experiment the magnetization is fully reversed while, in the STXM-FMR measurement, microwave off corresponds to the fully perpendicular case. Therefore, $2A$ has to be taken when comparing with $\Delta\mu_{abs}$. Geometrical considerations yield the full opening angle of the precession cone corresponding to 2θ, therefore yielding:

$$sin(2\theta) = \frac{2A}{\Delta\mu_{abs}}. \tag{5}$$

Here, $2A$ is (680 ± 31)cm^{-1} and $\Delta\mu_{abs} \approx 200{,}000$ cm^{-1}, which yields an opening angle of $\theta = 0.10° \pm 0.01°$. As already pointed out before [15], one has to consider the effect of the pulse length of the X-rays on the measured intensity. Comparing the pulse length of 50 ps with the duration of a full precession cycle of 110.4 ps makes clear that each light pulse averages over a substantial fraction of the sine-like contrast variation in time which yields a reduction factor of 1.5 compared to an ideal sampling of the dynamic magnetic contrast. Therefore, the actual opening angle for this FMR measurement is $\theta = 0.15° \pm 0.02°$, which is of the same order as the previously reported opening angle of 0.1° for a Co-stripe [15]. It has to be taken into consideration that the obtained opening angle of the FMR is only the out-of-plane angle, which in turn can differ from the in plane angle due to the magnetic anisotropy of the thin film sample. In addition, the measured opening angle naturally depends also on the microwave power applied to the magnetic specimen. This cannot be measured directly and it differs from sample to sample since the contact from the standard SMA cabling to the lithographically fabricated microresonator are so far not perfectly impedance matched and thus the entire system can have different transmissivity/reflectivity for microwaves leading to a variation of the microwave power at the sample for different specimen.

4.3. Origin of the Background Signal

The origin of the background oscillation visible in Figure 2 IIa needs to be discussed. One has to keep in mind that the output signal of the avalanche photo diode is amplified by a factor of 1000 (60 dB) to be detected. Therefore, it is very sensitive to issues with the pre-amplification. The cables inside the STXM (power supply for the APD and signal output of the APD) can act as antennas for standing waves generated by the microwave excitation of the sample. This can cause false positives/negatives depending of the phase of the microwave with regard to the photon arrival time, which can be misinterpreted as bulk (low spatial frequency) dynamics. This is an issue since microvolts of induced voltage by the microwave can be amplified to a "photon" level in the signal output of the APD. Furthermore, common detection methods can only detect one photon per bunch. Multi photon events only register as single events. This creates a nonlinear, detector response that gets more pronounced for higher count rates, and can interfere with normalization of dynamic contrast when imaging samples with big static contrast. While the signal in dark areas (magnetic specimen) is linear, the signal in bright areas (SiN-membrane) is compressed, thus the normalization algorithms that work by averaging obtain a skewed response that can create false dynamic contrast proportional to the static contrast. However, if the dynamic contrast reverses when switching the helicity of the light, i.e., the fitted phase between a STXM-FMR measurement with σ^+ and σ^- is 180°, it can be concluded that the observed signal is a consequence of the dynamic magnetic response of the system according to the XMCD effect and thus the dynamic contrast is of magnetic nature and thus stems from the external excitation of the magnetization generated by the microwave. Before the contrast reversal is demonstrated experimentally, the physical meaning of the phase of the sine fit shall be addressed.

4.4. Absolute vs. Relative Phase

The absolute phase should be measured between the precessing magnetization and the arrival of the X-ray pulse. However, this is complicated due to several issues. First, as in any resonance experiment, there is a phase difference between the microwave excitation and the precessing magnetization. Second, the phase of the X-ray pulse with respect to the microwave excitation cannot be determined directly. For the synchronization between microwave and X-ray pulse, only the driving frequency of the rf-cavity is available. This frequency determines the internal time structure of the synchrotron beam, i.e., it splits the electron beam into individual bunches. Inside the undulator, each of the bunches emits an X-ray pulse of fixed duration. Therefore, the travel time of the electron bunches from the cavity to the undulator as well as the travel time of the X-ray pulse from the undulator to the sample have to be taken into account. These are in principle known and should be fixed values for a given synchrotron. In addition, the length of the used cabling has an influence on the phase as well and this changes when the microwave set-up including sample and micro-resonator is physically changed. In a practical experiment, the fitted phase φ is only a relative number and comprises all the above factors. Therefore, it can only be compared as long as the entire microwave set-up as well as the excitation frequency of the STXM-FMR experiment is not changed. This implies that it is only possible to compare relative phases within the same sample and not between different samples. In other words, the obtained phase $\varphi = -39° \pm 5°$ is basically meaningless for comparing different sample measured in the STXM-FMR. However, the relative phase of different measurements using the same parameters and sample upon, e.g., the reversal of the helicity of the light can be compared and—according to the XMCD effect—should be 180°.

5. Experimental Verification of the Magnetic Nature of the Dynamic Contrast

Having discussed the small variation of the absorption coefficient during precession with a small opening angle together with the presence of a rather pronounced background signal of the SiN-membrane, it is important to investigate the behavior of the dynamic magnetic contrast upon reversal of the helicity of the light to assure that a true XMCD effect is indeed observed.

5.1. Contrast Reversal with Helicity

As a first test experiment, two STXM-FMR measurements at 9.61 GHz were done using σ^+ and σ^- polarized X-rays at the Ni L_3-edge of the horizontal Py stripe of the T sample shown in Figure 1b. In Figure 4a, the chemical contrast is shown while the individual six phases with σ^+ (blue) and σ^- (red) light are displayed in Figure 4b, top two rows. All contrast variations in (b) are shown on the same scale in order to emphasize the difference in $\Delta\mu \cdot t$ for the different measurements. Figure 4c collates the averaged dynamic magnetic contrast for all six phases derived by averaging over the respective marked areas. The RoIs were identified using the chemical contrast image in Figure 4a as indicated by the red box. The same colour scheme was used for the averaged intensities of the two measurements shown in Figure 4c.

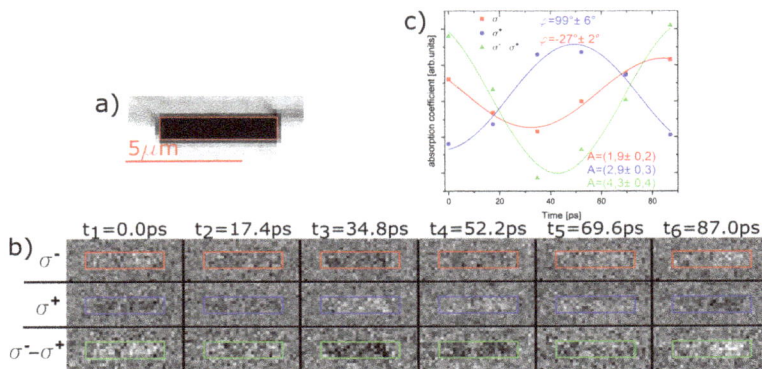

Figure 4. (**a**) chemical contrast picture of the Py stripe. In (**b**), two measurements of the dynamic magnetic contrast with different X-ray polarization taken at the Ni-L_3-edge are shown as well as the difference between the two measurements; (**c**) shows the average transmission intensity of the X-rays through the stripe sample for the three cases shown in (**b**). The averaged data was fitted with a microwave frequency of 9.61 GHz.

As one can clearly see in the individual phase images, there is a phase difference of $-27° \pm 2°$ for σ^- and $+99° \pm 6°$ for σ^+ light. Note that the value for the phases in Figures 3 and 4 differ because the samples and thus the microwave setup are different. The XMCD effect suggests that, by reversing the X-ray polarization, the relative phase should change by 180°, i.e., an ideal reversal of the contrast in all six images. However, the relative phase difference between the two measurement is only $127° \pm 8°$, which is significantly smaller. This is most likely due to the experimental constraint that only six phases can be resolved because of the X-ray pulse length of 50 ps, while the time difference between the individual phases is only 17.4 ps. Therefore, the experimental uncertainty is larger than the errors from the fitting procedure, especially considering the small overall size of the dynamic contrast change. In turn, one can increase the magnetic contrast by taking the difference between the two experiments with σ^+ and σ^- light according to usual XMCD experiments. The result is shown in Figure 4b,c (green) and it is obvious that the dynamic contrast is enhanced significantly. Nevertheless, as visible in Figure 4c, the amplitude is not increased by a factor of 2 as expected, which is due to the non-ideal reversal of the contrast as reflected by the behavior of the relative phases. Nevertheless, this is a first indication that a homogeneous FMR excitation behaves the same way as previously observed spin wave excitations [15].

5.2. Helicity versus Field Direction

In Figure 5 a second control STXM-FMR measurement done at the Maxymus endstation at BESSY II is shown where the STXM-FMR was measured with a slightly increased phase resolution of seven images for one full precession cycle. In addition to reversing the helicity of the light, one helicity was also measured for two different external magnetic field orientations B_{ext}, i.e., (σ^+, B^+), (σ^+, B^-), and (σ^-, B^+). The relative orientation of B_{ext} is indicated together with the image of the chemical contrast in Figure 5, top. The RoI where the contrast is spatially averaged is indicated by the red boxes in all images to ensure that the observed averaged signal only originates from the stripe and does not contain the SiN-membrane background (see above). The time-normalized spatial average intensity for each phase of the three different measurements is shown in Figure 5. The measurement in a positive magnetic field B+ and circular polarization σ^+ is shown on the left side of Figure 5a. The resulting averaged normalized X-ray transmission can be found in Figure 5, where it was fitted using a sine function with the microwave frequency of 6.785 GHz. This fit yields a relative phase between the X-ray pulses and the magnetization precession of $102° \pm 12°$.

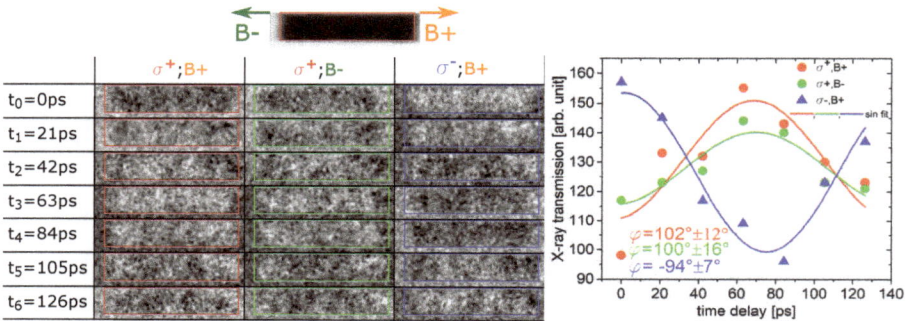

Figure 5. STXM-FMR measurement done at the Maxymus beamline at the Fe-L_3-edge at B+ = 60 mT and B- = −60 mT and different X-ray polarization σ^+ σ^-. The applied microwave frequency for all measurements shown in this figure was 6.785 GHz. The left-hand side shows the chemical contrast image together with the different directions for B- and B+. The averaged area is indicated by the red box. Below, the normalized intensity (with respect to the time average state) for the seven different excitation phases (or delay times) is shown for different field directions and X-ray polarisations. The spatially averaged intensity for each of these boxes can be found on the right-hand side with their respective colour coding.

As mentioned above, the sign of the static XMCD effect reverses when the external magnetic field along an axis of sensitivity is reversed [11,12]. At first glance, this would lead to a contrast reversal for the dynamic magnetic contrast in STXM-FMR as well. However, as can be seen in Figure 5a, the contrast does not reverse when the external field is reversed (middle column). This can be explained due to the fact that the present STXM-FMR configuration senses only the transversal dynamic component of the magnetization precession and not the direction of the magnetization itself. Due to the field reversal, the magnetization still precesses around the external field with the same phase relation as before. The dynamic magnetic contrast is only dependent on the projection of the dynamic magnetization onto the X-ray k-vector. This projection in turn exhibits a cosine behaviour and thus does not depend on the sense of rotation regarding the X-ray k-vector. This is evidenced by comparing the averaged dynamic magnetic contrast for σ^+, B^+ and σ^+, B^- in Figure 5. The resulting relative phase is $102° \pm 12°$ and $100° \pm 16°$, respectively, i.e., identical within error bars. In contrast, comparing σ^+ with σ^- polarization for B^+ in Figure 5 respectively, a clear contrast reversal is seen which is reflected by the resulting relative phases of $102° \pm 12°$ and $−94° \pm 7°$. The resulting phase change is thus $196° \pm 19°$ which agrees within error bars with the expected value of $180°$ for the ideal XMCD effect.

5.3. Contrast Reversal for Spin Wave Excitations

All previous examples were obtained with a uniform magnetic excitation, i.e., the uniform mode of FMR is excited. In these experiments, the oscillating contrast of the background was attributed to direct interactions between microwaves and the APD. Nonlinear APD responses can also have a non-negligible influence on the extraction of the dynamic magnetic contrast. This is especially relevant for homogeneous excitations of the magnetic specimen. In order to exclude this, the method of choice is an inhomogeneous excitation, i.e., a spin-wave of the microstripe is excited.

In Figure 6a, the STXM-FMR experiment, measured at the SSRL, of a spin wave excitation of the stripe parallel to the external magnetic field of a Py "T"-sample is shown. Details on these types of excitations go beyond the scope of this paper and will be discussed elsewhere; integral FMR measurements together with micromagnetic simulations have already identified these kind of spin wave excitations [3]. Figure 6a shows the measurement with σ^+ light, whereas, in b), the σ^- case can be seen. On the left-hand side, the chemical contrast images are provided while on the right-hand side a single image of the dynamic magnetic contrast is displayed. In order to maximize the contrast, the difference between the same two opposite phases has been taken for both polarizations. An additional smoothing as in Ref. [7] has been carried out to better visualize the inhomogeneous excitation. One can clearly see that there are regions with a pronounced magnetic contrast to either end of the stripe while the center shows a much weaker contrast with zero contrast in between. Importantly, the regions of strong magnetic contrast clearly reverse upon reversal of the helicity of the light while in other regions the contrast remains unaffected. Therefore, the contrast reversal is also observable with respect to a non-reversing region where the overall XA does not change, underlining that the contrast mechanism is indeed of magnetic origin.

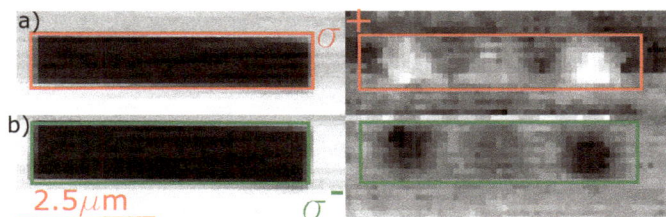

Figure 6. Comparison of the two circular polarizations for an inhomogeneous excitation of a T stripe. The left part shows the chemical contrast pictures for the different polarization measurements, respectively. For contrast maximization, the opposite phases of the same measurement are subtracted. The red and green boxes indicate the position of the Py stripe (extracted from the chemical contrast) for each of the two measurements to better visualize the excitation.

6. Conclusions

We have shown a way to correctly separate the quantitative pure dynamic magnetic contrast from the background signal in STXM-FMR. The opening angle of the FMR excitation of Py was determined at the Ni-L_3-edge by evaluating the amplitude of the dynamic magnetic contrast yielding an opening angle of 0.15° which corresponds well with previously reported order of magnitude for Co [15]. Furthermore, by switching the polarization of the X-ray photons from σ^+ to σ^-, the dynamic magnetic contrast switches its sign for the STXM-FMR measurement. However, contrary to static classical XMCD, a reversal of the external magnetic field does not change the dynamic magnetic contrast of the STXM-FMR because of the transversal geometry. Enhancement of the signal can be achieved by measuring the STXM-FMR with different polarizations (σ^+ and σ^-). Finally, the contrast reversal upon reversal of the helicity was observable for two different STXM-FMR setups, at two different synchrotrons, as well as for an inhomogeneous excitation. This makes evident that the contrast in STXM-FMR behaves similarly under reversal of the X-ray helicity to the static XMCD effect and one can take advantage of the unique combination of element selectivity and spatio-temporal resolution in future studies of magnetically excited micro- and nano-structures.

Author Contributions: Conceptualization, R.M., K.O. and A.N.; Data curation, T.S., T.F., S.P., M.B., D.S., V.N., S.W., M.W. and H.O.; Formal analysis, T.S., T.F. and S.P.; Funding acquisition, K.O. and A.N.; Investigation, T.S., T.F., S.P., R.M., M.B., D.S., V.N., S.W., M.W., H.O., K.O. and A.N.; Methodology, T.S., T.F., S.P., S.W., M.W., H.O., K.O. and A.N.; Project administration, K.O. and A.N.; Resources, M.F., H.W., S.W., M.W., H.O., K.O. and A.N.; Supervision, R.M., M.F., H.W., K.O. and A.N.; Visualization, T.S. and S.W.; Writing—original draft, T.S.; Writing—review and editing, T.F. and S.P., M.B., S.W., H.O., K.O. and A.N.

Funding: This research was funded the Austrian Science Foundation (FWF), project No. I-3050 as well as the German Research Foundation (DFG), project No. OL513/1-1.

Acknowledgments: Use of the Stanford Synchrotron Radiation Lightsource, SLAC National Accelerator Laboratory, is supported by the U.S. Department of Energy, Office of Science, Office of Basic Energy Sciences under Contract No. DE-AC02-76SF00515. Part of the measurements were carried out at the MAXYMUS endstation at BESSY II at the Helmholtz-Zentrum Berlin. We thank the Helmholtz Center Berlin for the allocation of synchrotron radiation beamtime.

Conflicts of Interest: The authors declare no conflict of interest.

References

1. Poole, C.P. *Electron Spin Resonance: A Comprehensive Treatise and Experimental Techniques*; Dover Publications Inc.: New York, NY, USA, 1997.

2. Narkowicz, R.; Suter, D.; Stonies, R. Planar microresonators for EPR experiments. *J. Magn. Reson.* **2005**, *175*, 275–284. [CrossRef] [PubMed]

3. Banholzer, A.; Narkowicz, R.; Hassel, C.; Meckenstock, R.; Stienen, S.; Posth, O.; Suter, D.; Farle, M.; Lindner, J. Visualization of spin dynamics in single nanosized magnetic elements. *Nanotechnology* **2011**, *22*, 295713. [CrossRef] [PubMed]

4. Rosner, B.T.; van der Weide, D.W. High-frequency near-field microscopy. *Rev. Sci. Instrum.* **2002**, *73*, 2505. [CrossRef]

5. Demokritov, S.; Hillebrands, B.; Slavin, A.N. Brillouin light scattering studies of confined spin waves: Linear and nonlinear confinement. *Phys. Rep.* **2001**, *348*, 441–489. [CrossRef]

6. Volodin, A.; Buntinx, D.; Brems, S.; Van Haesendonck, C. Piezoresistive detection-based ferromagnetic resonance force microscopy of microfabricated exchange bias systems. *Appl. Phys. Lett.* **2004**, *85*, 5935. [CrossRef]

7. Schaffers, T.; Meckenstock, R.; Spoddig, D.; Feggeler, T.; Ollefs, K.; Schöppner, C.; Bonetti, S.; Ohldag, H.; Farle, M.; Ney, A. The combination of micro-resonators with spatially resolved ferromagnetic resonance. *Rev. Sci. Instrum.* **2017**, *88*, 093703. [CrossRef] [PubMed]

8. Frömter, R.; Kloodt, F.; Rößler, S.; Frauen, A.; Staeck, P.; Cavicchia, D.R.; Bocklage, L.; Röbisch, V.; Quandt, E; Oepen, H.P. Time-resolved scanning electron microscopy with polarization analysis. *Appl. Phys. Lett.* **2016**, *108*, 142401. [CrossRef]

9. Cheng, X.M.; Keavney, D.J. Studies of nanomagnetism using synchrotron-based X-ray photoemission electron microscopy (X-PEEM). *Rep. Prog. Phys.* **2012**, *75*, 026501. [CrossRef] [PubMed]

10. Schütz, G.; Wagner, W.; Wilhelm, W.; Kienle, P.; Zeller, R.; Frahm, R.; Materlik, G. Absorption of circularly polarized X rays in iron. *Phys. Rev. Lett.* **1986**, *58*, 737–740. [CrossRef] [PubMed]

11. Stöhr, J. X-ray magnetic circular dichroism spectroscopy of transition metal thin films. *J. Electron Spectrosc. Relat. Phenom.* **1995**, *75*, 253–272. [CrossRef]

12. Dürr, H.A.; Eimüller, T.; Elmers, H.-J.; Eisebitt, S.; Farle, M.; Kuch, W.; Matthes, F.; Martins, M.; Mertins, H.C.; Oppeneer, P.M.; et al. A Closer Look Into Magnetism: Opportunities With Synchrotron Radiation. *IEEE Trans. Magnet.* **2009**, *45*, 15–57. [CrossRef]

13. Ollefs, K.; Meckenstock, R.; Spoddig, D.; Römer, F.M.; Hassel, C.; Schöppner, C.; Ney, V.; Farle, M.; Ney, A. Toward broad-band X-ray detected ferromagnetic resonance in longitudinal geometry. *J. Appl. Phys.* **2015**, *117*, 223906. [CrossRef]

14. Puzic, A.; Van Waeyenberge, V.; Chou, K.W.; Fischer, P.; Stoll, H.; Schütz, G.; Tyliszczak, T.; Rott, K.; Brückl, H.; Reiss, G.; et al. Spatially resolved ferromagnetic resonance: Imaging of ferromagnetic eigenmodes. *J. Appl. Phys.* **2005**, *97*, 10E704. [CrossRef]

15. Bonetti, S.; Kukreja, R.; Chen, Z.; Spoddig, D.; Ollefs, K.; Schöppner, C.; Meckenstock, R.; Ney, A.; Pinto, J.; Houanche, R.; et al. Microwave soft X-ray microscopy for nanoscale magnetization dynamics in the 5–10 GHz frequency range. *Rev. Sci. Instrum.* **2015**, *86*, 093703. [CrossRef] [PubMed]

16. Wintz, S.; Tiberkevich, V.; Weigand, M.; Raabe, J.; Linder, J.; Erbe, A.; Slavin, A.; Fassbender, J. Magnetic vortex cores as tunable spin-wave emitters. *Nat. Nanotechnol.* **2016**, *11*, 948–953. [CrossRef] [PubMed]

17. Feggeler, T.; Meckenstock, R.; Spoddig, D.; Schöppner, C.; Zingsem, B.; Schaffers, T.; Pile, S.; Ohldag, H.; Wende, H.; Farle, M.; et al. Direct visualization of dynamic magnetic coupling in a Co/Py double layer with ps and nm resolution. *arXiv* **2019**, arXiv:1905.06772.

18. Kukreja, R.; Bonetti, S.; Chen, Z.; Backes, D.; Acremann, Y.; Katine, J.A.; Kent, A.D.; Dürr, H.A.; Ohldag, H.; Stöhr, J. X-ray Detection of Transient Magnetic Moments Induced by a Spin Current in Cu. *Phys. Rev. Lett.* **2015**, *115*, 096601. [CrossRef] [PubMed]

19. Stein, F.U.; Bocklage, L.; Weigand, M.; Meier, G. Time-resolved imaging of nonlinear magnetic domain-wall dynamics in ferromagnetic nanowires. *Sci. Rep.* **2013**, *3*, 1737. [CrossRef]

20. Weigand, M. Realization of a new Magnetic Scanning X-ray Microscope and Investigation of Landau Structures under Pulsed Field Excitation. Ph.D. Thesis, Cuvillier Verlag, Göttingen, Germany, 2014.

21. Swinehart, D.F. The Beer-Lambert Law. *J. Chem. Educ.* **1962**, *39*, 333. [CrossRef]

MDPI

St. Alban-Anlage 66

4052 Basel

Switzerland

Tel. +41 61 683 77 34

Fax +41 61 302 89 18

www.mdpi.com

Nanomaterials Editorial Office

E-mail: nanomaterials@mdpi.com

www.mdpi.com/journal/nanomaterials

www.ingramcontent.com/pod-product-compliance
Lightning Source LLC
Chambersburg PA
CBHW051910210326
41597CB00033B/6093